Praise for *Sul*

"Holmes has that rare ability to ta... subjects and not only make them endlessly fascinating to read about, but also funny as hell . . . Reading Holmes' account is a bit like watching an awesome National Geographic documentary if it were hosted by a nature-loving version of Sarah Vowell."
—Debbie Stoller, *Bust* **magazine**

"Captivating . . . Holmes . . . deserves her own TV show. It could be called *Suburban Safari*, just like her newly published book, and could feature footage of Holmes performing all the exciting activities she recounts in print: chasing ants, scooping up slugs, setting traps with peanut butter and spying on spiders . . . A frolicking, rollicking romp of a book."*—Atlanta Journal-Constitution*

"In her own little kingdom, Holmes witnesses dramas one would sooner associate with the woolliest jungle or savannah, from chemical warfare between rival plants to the bloody mating duels of male squirrels."*—Forbes*

"Holmes reveals the intriguing, unnoticed dramas that unfold daily on suburban lawns. The book features birds, small animals and plant life immersed in a real-life plot that ranges from suspenseful to surprising, from sad to joyful . . . A fascinating journey."*—Denver Post*

"Entertaining and effortlessly compelling . . . What drives Holmes is not just concern for the natural environment but a ravenous curiosity about every aspect of the world around her, from the sex lives of dragonflies and squirrels, to the murderous tendencies of the English sparrows that have colonized her land . . . An intimate, wry, and often challenging look at a world most of us never bother to notice."*—Amazon.com*

"Everything about *Suburban Safari* is refreshing: its exploration of the nature just outside our back doors; Holmes's deft excursions into climatology, geology, and other -ologies; and a whole pile of eyebrow-raising facts . . . You'll finish the book knowing more about your immediate surroundings than you ever thought you wanted to—and it'll be the most entertaining education you ever get."—*Portland Phoenix*

"Holmes manages to find signs of hope and humor amid the spread of civilization, and she reports animal activities in her yard with the fervor of Wild Kingdom's Marlin Perkins and the laconic glee of Garrison Keillor . . . By the end of her year, Holmes has gently taught us that the American lawn is a pesticide-laden patchwork that's increasing by a million acres every year, that heating a house can produce five tons of pollutants annually and that stewardship of our own backyards is our responsibility."—*Publishers Weekly*

"With infectious enthusiasm and faith in nature's doggedness in the face of encroaching humanity, science writer Holmes follows the four seasons as they play out in her own micro-habitat . . . A cracking good reminder that an appreciation of the wonders of nature need not be reserved for special occasions."—*Kirkus Reviews*

"This is not just a very funny and very informative piece of writing, and not just a squirrel's horde of interesting information about the place you live. It's also a very important book—a graceful and forceful reminder that the natural world is everywhere all around us, to be savored and to be protected."—**Bill McKibben, author of** *Enough* **and** *The End of Nature*

"Look not to the faraway and exotic locale for the species-destroying and biologically undiversified mess we've gotten ourselves into. Look in your own backyard—or Hannah Holmes's backyard, where, with reverent wonder, she looks hard at her own soils, slugs, and sowbugs to show us the grand implications of the tiniest lawn-mowing decisions. *Suburban Safari* proves once and for all that there is life in the suburbs and that it's worth thinking hard about how to handle it."—**Robert Sullivan, bestselling author of** *Rats*

SUBURBAN SAFARI

A YEAR ON THE LAWN

HANNAH HOLMES

BLOOMSBURY

Published by Bloomsbury Publishing, New York and London
Distributed to the trade by Holtzbrinck Publishers

All papers used by Bloomsbury Publishing are natural, recyclable products made from wood grown in well-managed forests. The manufacturing processes conform to the environmental regulations of the country of origin.

The Library of Congress has cataloged the hardcover edition as follows:

Holmes, Hannah, 1963–
    Suburban safari : a year on the lawn / Hannah Holmes.—1st U.S. ed.
        p. cm.
    Includes bibliographical references.
    ISBN 1-58234-479-5 (hc)
    1. Urban animals—Maine. 2. Household animals—Maine. 3. Urban ecology—Maine.
I. Title.

    QL181.H65 2005
    577.5'6'09741—dc22

                                                                    2004017804

First published in the United States by Bloomsbury Publishing in 2005
This paperback edition published in 2006

ISBN-10: 1-59691-091-7
ISBN-13: 978-1-59691-091-1

1 3 5 7 9 10 8 6 4 2

Typeset by Palimpsest Book Production Limited, Polmont, Stirlingshire, Scotland
Printed in the United States of America by Quebecor World Fairfield

In memory of Cheeky

CONTENTS

INTRODUCTION

I GREW UP in a clan that disdained suburbs and cities. We were country folk. Our kind milked our own cow, boiled maple sap from our own trees, and raised pet robins and seagulls from eggs. We pitied city dwellers, who in the summer occasionally pulled into our driveway and pleaded with us to show their children how the carrots grew right out of the dirt. We were immodest about our ability to take care of ourselves in Nature.

But for me, going to school smelling of cow dung wore thin. I rebelled, running headlong for the city. For many years, I grew not one mouthful of food, nor even a rose. In the park, I didn't know one tree from the next. I forgot the names of the birds. And one day, reading in my office at an environmental magazine, I stumbled on a quiz. "Name five resident and five migratory birds in your area," the quiz prodded. "What direction do your winter storms come from? When is the next full moon? Exactly how far have you fallen, young lady?"

I had fallen far, far. And my ignorance of the animals and plants who share the planet has been on my conscience ever since.

But for me, the answer was not to get back to the land. There is no land that I could justifiably get back to. Every new house-lot displaces a family of woodpeckers, or a raccoon, or a tree that has invested two centuries in its climb toward sunlight. No, I had already concluded that the salvation of what little Nature is left on earth will depend on people staying in the cities and towns. (And in the interest of full disclosure, I could never go back to life removed from art galleries and top-shelf baking chocolate, anyway.)

As it happened, I recently settled on two tenths of an acre near the ocean in South Portland, Maine. Having spent a number of years writing about the natural wonders in such exotic locales as Madagascar and Mongolia, I thought it only fair to approach my new backyard with the same sense of discovery. Just because an organism is local doesn't mean it's simple. My oak tree, for instance, has the ability to adjust its acorn production to manipulate the squirrel population. And that an organism is common doesn't mean that it's unremarkable. The ubiquitous groundhog can spend nine months of the year in hibernation, without eating or drinking.

But where this book is concerned, I could have plonked myself down any old place. The creatures I befriended, and the dramas of competition, mating, and dying that unfolded before me, are available all around us, all the time. Nature, oblivious to the edict that cities are for people, is carrying on with her business almost as though we don't exist.

Ultimately, my reason for spending a year in the yard was not to acquire knowledge for knowledge's sake (although that's plenty amusing). Rather, it was to learn how to administer this patch of ground in the best interest of all its citizens: the chipmunks and the Baltimore orioles, the skunks, and even the humble mosses. In a world where humanity has climbed into the position of Biological Boss, I aim to become a benevolent dictator. But I can only rule fairly if I'm familiar with the needs and aversions of all my subjects.

As it turns out, taking care of Nature is a skill quite different from taking care of myself *in* Nature. That, in hindsight, was simple. Trying to balance the needs of all the organisms in my yard (myself included) is complicated. Everything is connected, for better or worse. Every step of my foot affects my neighbors. My smallest decision—to mow the grass weekly or biweekly, to cut down a tree, to fill the birdbath—rocks their world. When I trap mice in my cellar, I reduce the food supply to any owls who might contemplate making my yard their home. Feeding the local crow family allows me to observe their behavior and know them better, but it tilts their diet out of the natural range—perhaps shortening their lives? If I trim back the invasive bittersweet vines to make room for native sumac trees, the moss loses its shade and turns brown.

After a year in the yard, I can see that after a lifetime in the yard, I would still be insufficiently trained to administer this two-tenths-acre empire. But you have to start somewhere.

The Book

I chose to spend a year immersed in the lives of my out-back neighbors because it's a neat round of time. With the exception of my own breed, all the species who live here order their lives around the changing seasons. The plants blossom as the insects emerge from winter suspension to pollinate them. Migratory birds move north in synchrony with the emerging insects, fueling up on fresh meat wherever they stop. Squirrel mothers gestate through the hungry months of late winter so that their young will have time to stash away acorns before the snow flies again. In the fruiting season the crows use the bonanza of calories to grow fresh feathers that will see them through the insults of winter. In the short days of autumn, even the trees alter their cells in preparation for the freeze.

To my breed, which can purchase corn on the cob at any time of year, the seasons have lost some relevance. But to the crowd out back, the strengthening of the springtime sunlight and the ripening of late-summer berries retain a life-and-death importance. Because Nature still observes springtime as the season of new beginnings, that is where I began my observations, as well.

The scope of my investigation proved to be far too broad for a single soul to cover. So whenever I could, I enticed scientists—scholars of bugs, mice, or even weeds—to come by and shine the light of their expertise on my yard. This was an honor and a pleasure along the lines of Julia Child, James Beard, and Emeril dropping in to cook me dinner. To see my yard through such experienced eyes was akin to peeking through X-ray glasses.

3

My Lawn

What's a suburb? The word seems to mean something different to everyone. When I lived in a three-apartment brownstone in Brooklyn, New York, I felt as though I was in the suburbs of Manhattan. Now, in my neighborhood of tenth-acre lots, I feel ten times more suburban. Yet, to many minds, such a neighborhood is decidedly urban, and the five-acre and ten-acre lots farther from town constitute the real suburbs.

It's a moot point, really. If you define a neighborhood by the types of wildlife who make a living there, then large and small lots have much in common. Such "large-lot" creatures as moose, coyotes, and black bears take the occasional pass through my small-lot neighborhood. Even in the heart of the big city of Portland, a mile north of my yard, many of the same creatures who inhabit my yard earn their daily bread in the streets and trees: raccoons, squirrels, mockingbirds, skunks, mice, hawks, and oodles of insects. Recently, Portland police shot a moose who had roamed into the thick of things and was unlikely to roam back out without causing vehicular mayhem. Nature has never shown much regard for the lines we people draw on maps. Nature goes where Nature can.

I am surprised by the amount of Nature that exploits my yard. I had been led to believe that lawns are the abomination of the natural world. And perhaps this is so on the ideal lawn. The American lawn, as promoted by the lawn-care industry, is essentially a chemically supported grass farm. The plants themselves—lawn grasses—are not native to North America. They need more water than Nature provides in most parts of this continent. They're attacked by a plethora of insects, who must then be poisoned. Their ranks are also invaded by other plants, which engenders further chemical warfare. And unless they're mowed regularly, grass plants go to seed and turn brown.

The resulting grass farm is pretty to look at and fun to do somersaults upon. But it's not terribly welcoming to Nature. Even the occasional insect who can survive pesticides many not be able to make a living on grass. Many insects are particular about which plants they'll eat. And where insects are missing from the land-scape, other animals will be, too.

Fortuitously, most yards also host some shrubs and trees. This is the mixture that makes Nature happy. From a bug's-eye perspect-ive, or a bird's-eye view, most yards present a rich array of oppor-tunities: There's a "prairie" of lawn, a "savanna" of shrubs, and a "forest" of trees, all within a few flaps of the wing. Many animals thrive on this patchwork of habitats—many more than I would have guessed before I delved into this world.

Two important features set my own yard apart from the grass-farm ideal. One, it gets no chemicals. This results in a vibrant community of soil organisms and insects. That, in turn, attracts a community of insect-eating birds and beasts, and a great diversity of plants. Which brings me to the other feature. My lawn is not pure grass. It's a self-assembled community of grasses and weeds who can survive having their heads whacked off by the lawn mower every week or two. This diverse food supply also contributes to a diverse insect population, which attracts other animals. A less important feature of my lawn is that it's twice as long as the neigh-borhood average, comprising two tenth-acre house lots.

Other than that, it's a typical American lawn. It has a few flower beds, some ornamental shrubs, an oak tree, a couple of pines, a chokecherry tree, and some sumacs. Three craggy apple trees and a few stumps give evidence of the land's prior use, before my 1917 bungalow was built. The lawn is fenced with a haphazard array of pickets, galvanized fence, and rusty wire, but this doesn't seem to limit the traffic of skunks, possums, and other midsize animals. On one side, my yard is bounded by a neglected lot, which has been conquered by two invasive plants, Japanese knotweed (aka bamboo) and Asiatic bittersweet. On the other side lies a neatly mown lawn and the continuation of the old orchard.

When I began paying attention to the yard, I discovered a vast assortment of neighbors who had been flying under the radar, so to speak. Many town-dwelling animals are nocturnal, which had minimized my overlap with them. Others are subtle, such as the hawk who killed and consumed a bird in the yard one day. Had

5

I not seen the split second of arrival, I'd have missed the whole thing. And still others are so familiar that they had become invisible to me. On close examination, the squirrels and crows proved to be individual characters, each with a distinct personality, and all busy with work and play in my yard every day. It was heartening to find so much natural activity in such a peopled environment, and it was a pleasure to make the acquaintance of so many citizens in my domain.

The good news isn't unvarnished. Not every animal thrills to the sound of the lawn mower. That's unfortunate because new lawn is rolling out over the United States at a rate of one million acres a year. And even if all those acres were plugged full of a diversity of plants, and even if they were kept free of chemicals, some animals simply cannot adapt to life on the patchwork habitat we create. For instance, one of the marvels in my yard is the warbler migration that comes through in the spring. Warblers seem such treasures because they're fleeting, just resting here for a day or two. They can only nest and raise a family when they reach a wilder habitat. Every year, in my area and across North America, these birds have to fly longer and harder to find their niche. An even more daunting challenge faces a rare plant when its wild home is flattened: Unlike birds, plants don't have the option of flying onward. And plants are the anchors that hold down a food chain.

Besides erasing habitat and disrupting the food chain, new lawns also beget more lawn-mower fumes in the air, more pesticides in the water, and more traffic on the roads as people travel between their grass farm and their work. So the spreading of the lawn is a problem. The solution? Simply put, packing people into towns leaves more acres for wildlife and exposes fewer acres to our human clutter.

The very happy news is that we who crave greenery can find our solace on a fairly small patch of ground. From the core of the city to the edge of the forest, Nature is busy eating, growing, fighting, reproducing, dying. Absorbing the drama is the easiest thing on earth to do. All it takes is a lawn chair and a closer look.

SPRING

1

A FLOOD OF FEATHERS

I'VE ONLY BEEN investigating my backyard for a few weeks, and already much of what I've seen is weird. The other day, I discovered a strip of mummified squirrel skin jammed in the crotch of a sumac shrub. A few weeks ago, I found an English sparrow crammed headfirst into a tangle of forsythia branches, killer unknown. Today, April Fools' Day no less, crows are flying around wearing mustaches.

When I dedicated a year to watching my backyard, I suspected the average lawn hosts more activity than meets a casual glance. But I didn't expect to find stashed carcasses. Or to witness tail-pulling, or pear parties. And that's just the crows. Before long, the squirrels will prove just as wacky. Trees, even, will amaze me, due to the spite they direct at their enemies. But for now, it's the crows who have captured my interest. They're big, they're bold, and every time I settle in the yard, they do something bizarre. Today, it's the mustaches. The effect is created by many strips of maroon material draped in their beaks. Are they collecting something to make a nest? I doubt it. For one thing, there are four of them, and birds nest in pairs. Secondly, a crow's nest is a sink-sized mess of sticks, not a drapey, maroon basket.

As I'm pondering this, four crows without mustaches sail out of the sky and land on the grapevines that climb into my oak tree. Bracing their feet, they peck and pull off strips of peeling grape bark. When each has accumulated a mustache's worth, they all clamber forth from the vines and fly east. Later in the day, a solo crow lands on the brown grass of my lawn and beaks up a wedge of cold soil. She casts it aside and beaks up another. No good. She moves to my North Neighbor's yard and continues.

In short order I'm obsessed with my crows. I study their black faces through the binoculars as they swagger around the back of my lawn, prospecting for worms. After a few days of watching them, I can recognize each by its signature caw. If they're checking in with each other, one always says, "Caw, caw, caw!" Another says, "Caw, caw, caw, caw!" The third bobs slightly and says, "Cawwww . . . caw-caw!" The last says, "Cah-cah-cah!" In other parts of town, the crows I hear start to sound odd, foreign.

Reading about crows, I soon find Kevin McGowan, a Cornell ornithologist. At his Web site, I learn about a crow's favorite nest-lining material: maroon-colored grapevine bark. I learn what crows use to bind the sticks of their nest: gooey soil. I discover that my four crows are a family unit, composed of a breeding pair and two grown children. And I learn that this is my very own family of crows—or rather, that I'm one of their people. Each family of crows owns a bunch of blocks. My lawn may have been in their family for generations. We share a neighborhood, these crows and I. When I read that the West Nile virus is once again killing the crows of New York, it requires some effort to adopt a cool attitude of scientific curiosity: how interesting if West Nile should kill my crows! There are plenty of crows in the world, after all. Especially in the suburbs, especially in the past few decades.

If crows took their time cozying up to people, that's because our relationship with them was so nasty for so long. For centuries, crows filched grain and other food from people; and people responded with bird shot. Crows were slow to forget. But they're naturally adaptable. After it became a no-no to discharge firearms in the suburbs, the crows poked their heads in. And they saw paradise: lawns seething with worms; copious compost and garbage; and lighted parks and parking lots for sleeping. (In the nightmares of crows, I learn on McGowan's site, silent owls drop out of the blackness. They bite the heads off crows and swallow their warm breast meat. Crows like to sleep with the night-light on.)

And so the crows have come, and they have multiplied. A few weeks after the grapevine episode, my crows cease eating the worms

10

they pull from the lawn. Instead they carry them east. I follow. I'm looking for pine trees, based on two pieces of homemade data: One, in the top of my neighbor's fifty-foot white pine is a previous year's nest, shedding sticks. Two, the deciduous trees in the neighborhood are still puffs of green and pink, showing only nubbins of this year's leaves. I imagine a clever bird would choose the shelter and secrecy of an evergreen.

In a quiet lane a father and daughter bounce a ball. Across the road stands a row of pines. When I ask if they've seen crows in the trees, Dad says, "All the time." I squint into the treetops, seeing nothing. Then a seagull drifts near the pines, and a crow emerges. He strikes out after a seagull, muttering in flight. The gull lazily evades and drifts back for another peek at the nest. The crow flaps beside the gull, holding it at a distance. Three more seagulls spiral in. The crow flaps harder. Like a cutting horse, he wheels and chases, wheels and chases, muttering. The seagulls slide away on the wind.

The nest from which the crow bolted is motionless except for the tossing of the wind. When I look away from it, I have a hard time finding it again. That's what the parents were aiming for. If a raccoon or seagull gets to their eggs or hatchlings, they'll have to choose a new tree, build a new nest, and start incubating all over again. It's a tough business. Fewer than half of crow couples succeed in raising children in any given year.

I've been watching the nest for half an hour before I spot a lone crow in a tree a hundred feet from the nest. After a few minutes, she moves to another tree. Then another. In the coming days as I visit the nest, I'll find that crows take this sentry job seriously. One or two usually stand watch near the nest. When the family goes worming together, one remains in a tree, surveying the scene. My admiration for crow intellect will grow, too, as I watch them. Schoolchildren walking home past the nest are no cause for caws from a sentry. But I, pointing my binoculars at the nest, certainly am. Sometimes a sentry sneaks up behind me and sits overhead. Then if Pa Crow flies toward the nest carrying food for Ma Crow, the sentry above me barks the heads-up call:

11

"Cot–cot–cot–cot–cot!" Pa Crow veers off and lands elsewhere.

But even with my meddling, the crows have it easy. The suburbs are so bountiful that a family of town crows can make a living in a territory one half or even one third the size needed by their country cousins.

When I tear my attention from the crows, I learn that the suburbs are just as welcoming to many other birds. With a variety of landscapes offering shelter, food, and nest sites, and with people supplementing food and water, many bird species are content to share space with people. Bird surveys confirm this over and over: To find the greatest number of birds, get yourself out of the woods and into the suburbs.

Mind you, on close inspection, the bird bonanza may not feature your favorite species. For instance, one way scientists measure birdiness is in "biomass," which is pounds of life per square mile. When you measure birds in tons, they prove to be rare in rural areas, more plentiful in the suburbs, and thickest in the heart of the city. But biomass isn't necessarily beautiful. Most of that downtown bird bonus comes in the form of hefty pigeons. The situation in the suburbs is similar. Although the suburbs attract many species, native birds are underrepresented. Plentiful are European starlings, English sparrows (aka house sparrows), and those natives like cowbirds, cardinals, and chickadees, who can make a living in the mowed-grass-and-shade-tree ecosystem we've invented. Absent (normally) are those indigenous birds that need a deep forest, a tall-grass field, or a babbling brook.

Just at the moment, though, abnormality is in fine feather. A wave of plumage is breaking over the northern half of the entire continent. On their way to summer breeding grounds, birds of all stripes are pausing in my yard to tank up on the spring flush of insects, weed seeds, and nectar.

In May when my quince bush is a cloud of red blossoms, I sit on the porch each morning, waiting for the hummingbirds to arrive.

One bright day, they do. Ruby-throated hummers, they sway and glisten like drops of heavy liquid. The males' red bibs blaze in the sun.

Hummingbirds schedule their northerly migration to coincide with the opening of gas stations. In February they buzz away from their winter homes in the Caribbean and Central America. Some fly as far as five hundred miles in one shot, crossing the Gulf of Mexico. Others follow the land. Whenever they can, they embed themselves in a south-to-north wave of awakening flowers. They arrive at my quince when it's in full bloom. Now they can visit many of their daily five hundred to a thousand flowers without wasting gas commuting from one bush to another.

Hummers are ragged-edge birds. Like weight lifters, they're all muscle, and muscle burns food like crazy. They're bad at making fat, so they can't store much energy. Worse still, they lack the layer of downy feathers that keep other birds warm at night. But no animal can survive long on the ragged edge, so hummers have evolved their own way of saving calories: They hibernate for one night at a time. They routinely chill their bodies by four to eight degrees overnight, and if need be, they can turn the thermostat down sixteen degrees. Like tiny bears, they slow their breathing and depress their heart rate from 1,000 beats per minute to just 150. On a cold spring morning like this, my hummers may need fifteen minutes to power up to flying temperature.

Right now, they're awake. A little too awake. Two males and a female are mining the quince, and they're not sharing peaceably. They point their spears at each other and make blurry charges. For a small animal, a hummer can produce an intimidating series of twitters. Neighbor Hugh, walking by on the sidewalk, raises a hand to his face when he sees the birds sparring. "Whoa," he says. "Watch your eyes." I fill a feeder and hang it near the back deck. The hummers find it within seconds. One male declares it his private property. Between drinking binges, he stares at it from the depths of a forsythia bush. When I walk between him and his sugar water, he cusses. He buzzes hummers who approach his feeder, boys and girls alike. He seems to convert sugar water directly to belligerence. His aggression will probably win him a girl, whom he'll impregnate and abandon, leaving her to raise the children.

Every morning now, a new bird chimes in with the chorus. About eighty percent of the birds that breed in North America, including the orioles, goldfinches, and warblers that are brightening my muddy yard, have to find their way from points south to points north every spring. Some return to the same corner of the same yard where they nested the previous year.

How do they do it? Reading up on bird navigation, I stumble on this peculiar fact: A robin can find his way around with his left eye taped shut, but not with his right eye taped shut. The marvelous explanation offered by the researchers is that the earth's magnetic fields spark a chemical reaction inside the bird's right eyeball, transferring directional data to the brain. The magnetic information amounts to a map. Because magnetic fields shift and distort the map, birds can also see a plane of polarized light that follows the sun across the sky. They use the sun to correct the magnetic map. Songbirds, who migrate at night, use the stars to correct their map.

Eye-taping is the tip of the iceberg, in terms of migration experiments. Scientists also put birds in planetariums and spin the stars at abnormal speeds, to see if the birds reorient themselves accordingly. (They do.) They put impressionable young birds in planetariums with fake starscapes to see whether the star map is innate or learned. (It's learned.) They generate phony magnetic fields around birds, to see if their sense of direction warps accordingly. (It does.)

The birds winging into my yard this spring are the winners in the orienteering challenge. Untold numbers of losers exhausted themselves flying around lighted towers, got lost, blew off course, ran out of fuel over the ocean, or crossed the ocean to discover that a drought has killed their food.

Which brings me back to sustenance. I hung a hummingbird feeder because I thought these travelers needed extra calories. But of course, hummingbirds did just fine before ArtLine made feeders. And they evidently do just fine if you don't change the sugar water in the feeder. I didn't know that fungal spores can get in through the yellow flowers and sponsor dynasties in the syrup. After a couple of weeks, I don't see any more hummers. I figure they've migrated onward, but then I spy them jousting over the flower beds. When I wash the feeder and refill it, they return in a matter of hours. But they've obviously survived without me.

That hummers find a feeder so quickly hints at their smarts. Experiments show that they're not born with an urge to find red blooms. They learn this preference. When my hummer child eventually emerges from his half-walnut of a nest this summer, he'll display a frightful ignorance of botany. He'll hover hungrily in front of brown lilac seeds, tapping them with his needle. He'll investigate twigs. He'll probe leaves. But over time, his sugar-sensitive brain will identify a pattern: Many of the flowers with the highest sugar content are, by chance, red. Thus hummers learn to love red. They learn it so well that they'll sip sugar water from a red flower painted around a belly button, or, allegedly, from lipsticked and puckered lips.

❧

Another sugar junkie is stoking his mojo as he passes through the neighborhood. A Baltimore oriole, feathered in glowing orange, black, and white, perches on the post that holds the hummingbird feeder. I make a special trip to the supermarket for an orange and lay oriole-colored slices on the deck railing. But this guy has other plans.

I can't fathom the goal at the end of his plans, but this is how he carries them out: He hops around in one of my old apple trees, peeking into the clusters of emerging leaves. He spends hours doing this. Another male arrives and does the same. Separately, over three days, they examine every inch of each old apple tree in the neighborhood. They ignore the two pear trees. A week or so later when the trees blossom, the oriole plan gets a little clearer. Each apple tree unfolds only about a half dozen flowers. The orioles ate all the buds.

My botanical aunt tells me that this is completely normal for an oriole, but I fail to find printed proof in an academic journal. The oriole expert I locate tells me the birds simply wouldn't eat the buds, because they can't digest cellulose. But a friend checks her *Birder's Handbook* for me. Lo and behold, according to Paul Ehrlich and friends, orioles do indeed eat apple blossoms.

Now I'm intrigued. If the buds are made of indigestible cellulose, what makes them worth eating? After some poking around, I conclude it's their pink pigments. Through much of the bird world, it's the gentlemen, not the ladies, who dress in showy plumage. Ladies, according to the dominant theory, prefer the gent who wears the brightest colors. Birds can produce blue and green pigments in their own bodies. But birds with red, orange, or yellow plumage have to eat foods with carotenoid pigments in them. Male orioles fall into this category. And apple blossoms contain red pigment. Hence it is my personal theory that orioles eat blossoms to extract the carotenes. The only fly in my theoretical ointment is that birds can't store the pigments and have to eat them during the molt when they're growing new feathers. And the main molt comes in autumn. However, my oriole expert did let slip that some young males molt in the spring. So perhaps that was the ulterior oriole goal: The young men, eager to awe the ladies with their brilliance, strip-mined the apple trees for their colors.

Carotene coloring is popular with the birds in my yard. The cardinal and goldfinch fellows both are intensely colored from beak to butt. Biologists talk about animal features having "costs," because the labor required to build and maintain any one feature leaves less energy for other chores. A peacock's tail, for instance, is expensive because it's heavy to carry around, and it probably slows a peacock's escape from predators. The reddest cardinals have to locate a good source of red or yellow fruit and spend energy battling other males who also want it. If that red food isn't very nutritious, then the red feathers are even more expensive. Those pricey feathers must yield a benefit.

The reddest cardinal males, I learn, win the thickest, most predator-proof nest sites. That makes redness a "true signal," a reliable sign that the guy is fit. Further evidence that color indicates vitality comes from experiments with blackbirds, whose yellow bills fade when they're sick. Not surprisingly then, the reddest cardinals also attract the fittest females, those who start laying eggs early in the season, enabling them to raise two families by summer's end.

The day of winning a mate, however, is still distant for many of the birds in my yard. Most are only resting here before racing onward to stake a claim in their preferred habitat. Watching the traffic at my rest area, I spot birds I've never seen before: a Swainson's thrush, a bewilderment of warblers, a white-crowned sparrow, a cuckoo, large and handsome flickers, Carolina wrens, and something I record in my notes as "a cute little leaf-kicker." Most of these will raise their families somewhere a little more private than the suburbs. The thrush wants a spruce forest, my bird guide says. The white-crowned sparrow is hoping for "boreal scrub." Heaven only knows what the leaf-kicker has in mind.

And heaven only knows where they'll all find their dream homes. One of the reasons the songbird population is crashing is that North America is being converted from spruce forest and boreal scrub into Wal-Marts and White-Crowned Sparrow Estates. Another reason is that the winter homes of many of these birds, in Mexico, Central America, and the Caribbean, are also being converted from bird habitat to habitat for people, cows, and sun-grown coffee trees. Most birds have evolved to eat specific seeds and insects; to use a particular plant for nest-making; and to build those nests in a certain density of forest to protect it from egg-stealing predators. When people turn a forest into a lawn, pasture, or parking lot, the resident birds must find another source of food, shelter, and home construction. They're failing with flying colors. It takes a special bird to deny millions of years of tradition and adapt its lifestyle to a new ecosystem. It takes a crow. A goose. Or a chickadee.

Mating season comes at different times for different birds. Today, it's the State Bird who's screaming bloody murder. One black-capped chickadee perches in my chokecherry tree, and another sits in the lilac hedge fifteen feet away. Once upon a time, I would have regarded their noise as "singing." But there's no joy in the hearts of these animals. They're hurling death threats across the lawn.

Well, it's not quite that bad. Their contest does have serious consequences, though. The winner will keep his girl at home. The loser's girl may wait until his back is turned, then flit to the challenger's branch. After a speedy rendezvous, she'll return to her partner. She'll lay an egg fertilized by the competitor. And her mate will squander his energy protecting and feeding another man's progeny.

This all hinges on a fellow's singing. The dueling is a daily affair, and the tune is a simple *fee-bee*. The average chickadee might *fee-bee* just to assert his property rights. But an aggressive chickadee will make a vocal assault. He'll adjust his pitch to match his opponent's, then try to drown him out. To the victor go the girls.

In recent years, scientists have concluded that many birds, males and females alike, are terrible cheats, obsessed with securing the best available genetic input for their offspring. But chickadees are particularly status conscious. The wife of my neighborhood's highest-ranking male is the woman most likely to cheat if she hears her husband shouted down. And if that woman should fall to a cat or a car, the second-ranked wife will dump her own husband to shack up with the bereaved. The State Birds may be cute, but they're tireless screamers, and ruthless schemers.

Chickadees aren't doing all the shouting, of course. Male birds throughout the neighborhood are so awash in springtime testosterone that they can't keep their mouths shut. Cardinal boys holler at each other, *"Wheet! Wheet! Wheet! Wheet!"* Dueling catbirds string together imitations of chickadees, cardinals, and seagulls, plus species I suspect them of inventing. A starling runs through a repertoire of mimicry that includes the terrified shriek of a two-year-old child. This is so realistic that, after I ascertain there's no terrified child involved, I conduct a Web search on *starling*. I find the site of a woman who has taught two rescued starlings to talk. I click a button. "Hello, Storm," Stormy says in a high and creepy whisper. "Stormy's a pretty bird."

The starling's repertoire, like the chickadee's volume, is a personal ad promoting a man's strength. In my North Neighbor's yard, a starling male stands beside the apple-tree hole he has procured and sings the longest, most complicated song he can muster. He warbles and clicks, creaks and tweedles, he screams like a toddler. Among nearby female starlings, the song is cause for speculation regarding the singer's health and intelligence. Near the end of his song, the soloist hunches over and twitches his wings as though he's being electrocuted. The girls melt. They flutter near to inspect the hole. The lucky birdie who moves in will find that her starling darling brings home flowers, even after they have a nestful of children. Well, not flowers exactly, more like herbs: wild carrot, yarrow, goldenrod, and fleabane, all of which are highly aromatic, and some of which have proven to impede the hatching of parasites that lurk in nests, menacing babies. A girl could do worse than to land a starling.

As the ground softens in the sun, a summer community takes shape. Behind the unfurling leaves of the oak and lilac, chokecherry and honeysuckle, and in the depths of the South Neighbor's Bamboo Wilderness, birds couple up. The slate-colored catbirds claim a back corner of the yard, and the starlings settle into their apple tree. With eggs accruing in pouches of sticks, weeds, and feathers throughout the neighborhood, the hormonal tide recedes, and the yelling does, too. "Songs" diminish and "calls" rise. Instead of *fee-bee,* chickadee boys and girls alike *chick-a-dee,* keeping in touch as they hunt insects on the sumac trees or tuck sunflower seeds behind my gutters. The cardinal couple exchange a domestic *pip.* Mr. Catbird lays off satirizing his neighbors and now swaps *mews* with his wife. If I wander too near his corner, he hops onto the fence and cries at me, *"Meeewwwwww!"* I cry back, and he cocks his head before repeating his accusation.

This is a high-stakes time for all the birds. The crows, like everyone else, have heavily invested in the contents of their nests. They burned countless hours and calories collecting sticks and mud and grapevine bark. Pa Crow led the family force to keep the territory free of invaders. Ma Crow channeled calories and nutrients into her eggs. Pa and the kids fed her as she sat for nearly three weeks, converting worms, roadkill squirrel, and pizza crusts into heat, which passed through the skin of her featherless brood-patch and warmed the eggs. If the nest tumbles in a high wind, or if a raccoon or squirrel gets the eggs, all that effort is wasted. But turnabout is fair play, I suppose. Crows themselves are famous home-wreckers. In addition to eating other birds' eggs, they'll kill adult birds. They'll even take on larger mammals. Lawrence Kilham, author of *The American Crow and the Common Raven,* reports watching crows hammer on a sickly wild pig and jab a newborn fawn. (The fawn was fine, but my admiration for crows felt the blow.)

One day at the end of May I follow a worm-bearing crow to the nest, and, Aha! Two expensive little heads protrude above the nest rim. An adult is gliding in to feed them. Her feet barely touch the nest. The heads rise as she approaches. Soundlessly, she tucks food into the open mouths. Then she's away from the tree, and the heads withdraw into the nest.

All the adult crows, both the parents and the elder siblings, bring food to the kids. It's unusual behavior, this sibling helpfulness. And apparently it's not all that helpful. Researchers have set up studies where they monitor some breeding couples who have helpers, and some with no helpers. There's no glaring difference in nesting success. Unassisted parents lose no more eggs or babies to predators, so the helpers aren't keeping the nest safer. Nor are more babies lost to starvation, so the food that helpers bring isn't crucial. The best guess is that the helpers are helping themselves: When they eventually have children, they'll be seasoned parents.

Then again, perhaps after these young helpers get well acquinted with their baby brother, they'll decide against reproducing. I name Yawp in early June, shortly after waking to a symphony of braying crows. I peek outside. The four adults are stationed around a neighbor's house, perched in trees and power lines. They're flinging gusts of complaint toward the neighbor's porch roof. On that roof is a crow, picking bits of gravel off the shingles, and toying with maple twirlies. From time to time, this crow looks around, opens his wide beak, and says, "Yawp." There's a hint of complaint in that expression. In tone, it's akin to a kazoo. When an adult flies overhead, the youngster flutters and gapes. "Yawwwwwwwwp!" The adults rain another storm of cawing upon him. Yawp picks up a maple seed and drops it. He has tried flying. He didn't care for it.

When I study up on this stage of crow life, I find that youngsters leave the nest a bit prematurely. They typically spend a few days on the ground or in low trees, figuring out the flying business. The parents feed them and worry loudly as the kids fumble around the neighborhood, dodging cats, dogs, cars, and crow haters. I go out to check on Yawp an hour after his debut and find him settled into the porch-roof gutter. The adults sulk nearby. Later, a neighbor tells me Yawp did finally make a move. He plopped down to the lawn, giving his family fits.

In the afternoon when I don't see him around, I walk to the nest. Both kids are back in it, but just barely. They're hopping from one side of the nest to the other, swapping places. Teetering on the edge, they lift their wings and flap. One crow flap-hops out onto a limb and nibbles an overhead stick. They're busy, busy, busy.

The next day the other kid, the quiet kid, gets her flying lesson. The four adults fly through the neighborhood in a clump. Tagging behind, veering all over the skyway, is a fifth crow. When the grown-ups reach their territorial boundary, they turn left, but their protege does not. They race back and herd her to a tall tree with slim, vertical branches. The young crow flaps and grips a branch. The branch bends to a right angle. She flaps to stay upright. But the branch can't support her. She flaps free and tries another. Over and over. Even when she finds a sturdy branch, the tossing wind keeps her off-balance.

It's baby season all over. In Neighbor Hugh's yard, English

sparrow babies are fluttering and begging loudly in the grass. In my yard, they're investigating the litter under the lilac hedge, testing everything in their beaks. Petals are rejected. Twigs are rejected. Leaves are rejected. I don't know what they accept, but they're exploring at full tilt. On my walkway a youngster I can't identify has discovered an endless source of little protein packets: He waits beside an anthill. The crow kids are fully mobile within a few days, and they, too, work the yard. Well, the quiet one does. She has a droopy wing, perhaps from crash-landing in a tree. Like her elders, she walks quietly through the grass, hunting worms. Droopy, I call her. Droopy is a diligent crow.

Yawp, though, is a nightmare crow. I can hear him from one end of the neighborhood to the other: "Yawp! Yawwwwp!" He starts at dawn, and he goes until dusk. Today he rushes each family member in turn as they pull up worms. "Yawwwwwwwwwwp!" He assails his first victim, but that bird turns away and swallows fast. Yawp runs toward his sister when she scores, but she shreds and eats her worm alone. Then a third crow flies into the yard, and when the squawling brat approaches, this crow opens wide. Yawp dives into the black throat to gulp a meal. There follows a gorgeous minute of quiet as he swallows. Then he resumes kazooing. He's a bottomless pit, and he has no intention of filling himself. One insight afforded by Yawp's ever-gaping maw is that young crows have pink mouth linings, in contrast to their parents' black ones.

Baby birds can be difficult to distinguish from adults. Some wear a duller suit of feathers than their parents, and many sport a wedgier bill, but these can be subtle distinctions. My crow babies do have brown feathers, as opposed to black. But the brown is so dark I can only distinguish them in direct sunlight. Even that clue is being diluted as new, black feathers emerge. So if Droopy's wing isn't drooping and Yawp isn't yawping, I have to fall back on clues won through hours of study with the binoculars, such as the pink mouths and the fat beaks. They also lack the pronounced ridge over the eyes that gives the grown-ups their noble expression. In profile, the kids have smaller heads. Their tails come to a square end, while adult tails are rounded. The kids' legs are set closer together, which produces a less swaggering gait. And they're a hair smaller than adults. But all these traits are so hard to spot that, as with pigeons, I expect most people believe they've never laid eyes on a baby crow.

In spite of Yawp, I'm growing fond of my crows. I head into the yard every day with good intentions—I'll watch the ants, I'll study the trees—and the crows distract me. Yawp becomes a great fan of hopping. One day he spends ten minutes hopping up onto the neighbor's birdbath, then hopping down again. Sometimes he hops straight up in the air and comes down facing a new direction. Whenever another crow pecks the ground, he bounds across the lawn in a series of gallopy hops. He bounces at starlings and pigeons, sending them whirring away. He hops at squirrels, who hop right back at him, swinging a paw. He picks up a green pear the size of a walnut and careens toward Droopy, holding it high. Droopy walks away from her brother. Yawp drops the pear and hops over her, landing in her path.

Droopy's talents lie in another direction. She's a stick girl. In her spare time she sits in an old apple tree and breaks off twigs. She toys with each for a few minutes, then drops it and chooses another. This can be a dangerous time to be friends with a crow. Someone has written to crow scholar Kevin McGowan's Web site about crows repeatedly peeling the rubber off their windshield wipers. Textbook juvenile behavior, McGowan replies.

While young crows are extra pesky, crows of any age can be terrible teasers. Kilham, author of the crow book, once watched a crow flaunting a piece of "food" to lure a wild turkey into chasing him; after a good romp across the field, the crow revealed his prize: a chunk of cow pie. Tee-hee! Kilham writes that they're also tail-pullers, yanking the tails of vultures and otters to make them abandon food. This I saw once myself, when a seagull was hogging peanuts I had spread for my crows. Four crows circled the gull, and one in the back leaned forward to yank a tail feather. In a similar vein, I once watched two crows at my neighborhood beach mugging a seagull. The gull stood over a crab, with a crow posted four feet to either side. The gull seemed to fear that if he shook the crab apart, pieces might spin toward a crow. So he waited, clucking. The crab gathered its wits and walked away. The seagull snatched it back. The crows gazed out to sea, as though butter wouldn't melt in their beaks. The seagull clucked louder. The crab again excused itself and was again dragged back. And then the seagull could stand it no longer and charged the right-hand crow. The left-hand crow flew away with the crab by the time the gull reconsidered. The two crows did not share.

In mid-July it is suddenly Droopy's turn to be annoying. One balmy morning I step outside to the soprano racket of a crow losing her mind. From the depths of an apple tree comes a string of contradictions and pronouncements that add up to gibberish. By now, I am fluent in crow. And it sounds to me as though Droopy has suddenly learned to speak, but has no inkling what words mean. First comes a long "Caaawwwwwww!" Then imme-diately, "Cot-cot-cot-cot!" And then, "Caw, caw, caw." If I may tender a translation, this means roughly, "I'm Droopy and every-thing's fine. *Horrors, a crow-eating devil!* Hey, guys, I found food! I'm Droopy, *and here comes a demon!* There's food here." At the end of a verse Droopy flies to the neighbor's cherry tree and clambers around looking for fruit. She babbles on. "Cawwwwwww! Cot-cot! Grrrrrrrack. Graaaaack. Cot-cot-cot! Tuck. Tuck."

For weeks she wakes the neighborhood at dawn with a string of witticisms she's thought up overnight. Her family learns to ignore her. They don't fly in to investigate her announcements of feasts or fiends. She chatters to herself as she hunts worms. When a family member down the block issues his routine sentry call, she tosses off a perfect imitation of him. (Crows, like starlings, are fair mimics, even managing human speech.) She attempts a series of super-quick caws but ends with a strangled cough. The girl even looks demented, with leftover baby down poking out through her feathers, and a lump on her droopy shoulder.

The whole family is molting and revolting. And it's not just them. All the neighborhood birds are facing a change of feathers, which are designed to last through one year of flying, fighting, and bumping into things. With summer food plentiful and the demands of nesting past, birds can now afford to invest in fresh plumage. I find something new on the ground every day: sapphire blue jay feathers; glossy crow feathers; salmon-pink cardinal down.

Flying with missing feathers demands extra calories, but the timing is not accidental: My yard is stuffed with food. The raspberries at the back are ripening. The catbirds, who live near there, toss back whole fruits. When I pick a few myself, Mr. Catbird greets me with what I like to think is a friendly tone. The wild grapes that drape the oak tree aren't ripe, but both catbirds and cardinals are harvesting them. Too big to swallow, the grapes have to be carried to the ground for butchering. This alarms me, given the number of cats that roam the neighborhood, so I keep watch while my comrades eat. Acorns are thudding down, too. Squirrels and blue jays tuck the nuts into the lawn, arranging the grass over them.

Cot-cot-cot! Someone is in the top of a pear tree, yelling for the family. Soon all six crows are crawling around the branches, jabbing unripe pears. Hard fruit thumps on the lawn. One by one, each crow skewers a hard pear, then flies to Neighbor Hugh's house. I don't know what's so compelling about his roof, but they line up on it, each pecking a pear clasped in his or her feet. The occasional pear gets loose and rolls down the shingles. The crow either catches up with it or returns to the tree for another.

The pears teach me something about animal appetites. Clearly, the crows love pears. At least twice a day, some of the clan gather to eat them. Two birds might climb in the branches, while two others peck the fruit that falls. But after an hour of pearing, they'll drift to worming or disappear toward the pizza joint at the end of my street. They like a varied diet. There are exceptions, of course, foods too precious to leave uneaten or uncached. Peanuts, I discover, are irresistible.

One day, I toss unsalted peanuts out onto the lawn when a crow is watching. This crow alerts the others. *Cot-cot-cot!* The family assembles in the trees, and when I retreat to the deck, they descend.

After just a few of these peanutting sessions, the crows know the sound of peanuts rattling in the plastic jar. If I feel silly singing out, "Crow-crow," I can shake the jar to summon them. The first few times, they approach the food cautiously, hopping at the peanuts, then leaping back as though they'd landed on hot coals. I've read a similar account of ravens approaching an animal carcass, in Bernd Heinrich's *Ravens in Winter*. Heinrich speculated that these jumps help the birds confirm that the prey is really, truly dead, not lying in wait for ravens. In that light, it's pathetic to see the crows tiptoeing around as though my peanuts could rear up and bite them.

When I run out of peanuts, I try the crows on old bread, past-prime rib, and withering grapes. Novelty, though, is never amusing to a crow. Bread seems familiar to them, and they snatch it readily. But the old steak is so scary that they visit periodically all day, bouncing and retreating. A skunk or possum drags it away in the night. Grapes are taken after much leaping and alarm. Even after the crows have consumed many meals of freezer-burned chicken, a piece that lands standing upright on the lawn is disconcerting. Yawp circles the upright morsel for a full five minutes, crouching and feinting at it. He finally steals close enough to extend one foot and tip it over. Then it's safe to seize.

Whenever the crows come to dine, they swallow a few bites and carry the rest away to stash for later. Peanuts they hold in the back of their throat and line up their beak until they can't pick up one without dropping another. Saltines they pile five high on the ground, then lift in a stack. Chicken they shred and swallow until the main chunk is light enough to carry away. Loaded down, they flap off to cache. One might fly to a nearby pine, poking his peanuts between branches. Another is partial to a tangle of dead sumacs in the South Neighbor's Bamboo Wilderness. They often tuck their goodies into tufts of my lawn, rearranging the grass over the stash so adroitly that I can never find it. The only time I ever find a crow cache in the yard, it'll tickle the top of my head. Ducking under a sumac, I'll look up and see a strip of squirrel skin, a white bone protruding from the bottom, wedged into the

26

sumac crotch. With the meals I've been serving, I'm not surprised they haven't come back for that.

In this fat season, all the birds are in constant motion. And only a fraction of their busyness involves eating. They're forever on the lookout for predators. Mr. and Mrs. Catbird fly to the berry patch in stages, perching and scanning before they flit into the canes. Then they grab a quick berry and return to a higher perch. They scan again before they toss back the fruit. They seem never to relax. And with good reason.

One day a shredding sound comes through the open window. I glance out just in time to see a big bird blasting out of the lilac hedge. He has dived straight down on a gang of English sparrows. Empty-handed, he coasts up to the oak tree. He settles and looks around, his head swiveling over a ginger-speckled breast. With my bird book I sit at the window and exercise my frustration. Bird books generally present three portraits of each bird: adult male; adult female; juvenile. But the things that fly into my line of sight seem to be transsexual, semimature birds blown four thousand miles out of their normal range. Today, typically, Roger Tory Peterson presents nothing that looks like my hunter. I call a birder friend, who opens her *Sibley*. As the hunter scans the yard, we conclude that he could be a teenaged sharp-shinned hawk.

When the crows discover him, they hurtle toward him braying. He lifts off, cutting around the trees. A minute later he appears high in the sky. And as Peterson says a sharpie should, his silhouette goes flap-flap-flap-flap-gliiide. He's just a youngster, more brawn than brain. But he'll be back in my yard every week or so, and eventually I'll see him sink his talons into a meal.

I'm soon confronted with another threat to my birds. On a wet day when I go to pick berries, Mr. Catbird doesn't show up to *mew* at me. The little babbler, who spins off the songs of cardinals, seagulls, kiwis, and nitwits, is silent. Well, I think to myself, it's a chaotic time in the avian world. Kids are out and about,

27

terrifying their parents with their ignorance. In some species, families are merging into crowds who travel the neighborhood in division strength. Territories, so hard-won a few months ago, are disintegrating.

Returning to the house, I note a gray breast feather in the damp grass. "Molting," I think. Then I see a dark tail feather. I don't want to look for more. I make a methodical survey anyway. A downy feather and a gray wing feather lie near the deck. A big tail feather, broken and gritty, is near the street.

I hope to salvage some peace of mind from the possibility that the sharpie took my catbird. But this trail of feathers marks the highway that a half dozen neighborhood cats take through my yard. I can imagine my catbird beating his wings against cat and soil as he was carried away. Death-by-Sharpie I could have accepted. Hawks need to eat. But the cat had probably breakfasted on kitty kibbles. The cat probably killed my catbird solely at the urging of instinct. That's what I'll read later, along with a litany of additional sorrows: Bells are little help. Declawing is no cure. Overfeeding doesn't work. Cats are hardwired to hunt, whether they're hungry or not. And every year in the United States, they kill hundreds of millions, perhaps billions, of birds that are protected by federal law. Even birds that break free from cat claws risk a fatal infection.

I'll learn that later. This morning, though, I'll stand under the sumac, out of sight of the neighbors, and drop my head in my hands. In my grief, I envision a day when my yard is silent except for the meowing of other people's cats. I think about trapping the cats and tying notes to their collars for their owners to read. I think of calling the Justice Department to tell them that federal law bans the killing of catbirds, and to insist that every cat owner on the street be rounded up and questioned about this felony. The weather is gloomy, and that's my excuse for staying indoors the rest of the day. The truth is, the lawn feels like a grave. I wonder if it would help to gather up the feathers. I could bury them. Or put them in a hole in the apple tree. Or build miniature arrows from them, to shoot at cats.

The following day in the yard, this ain't no lie, bird motion overhead catches my attention. Pecking tentatively at the ripening grapes is a new-minted catbird, his tail and beak blunt with youth. When he flicks his tail, it's an abbreviated gesture. But he flicks it just the same, tilting his black eye to regard me. He flies to an apple tree and snatches a large insect. It's big enough that he has to toss it around to get it aimed down his throat. It squirms. Then he swallows. His voice is young and tentative, but with an eye on me he uses it: *Mew!*

A better woman than I would mourn the murdered insect, now being crushed in the grit and muscle of the catbird's gizzard. But we choose what we love. I love my catbirds.

Furthermore, I will risk the disdain of millions and confess that I am also coming to adore my crows. Later this summer, I'll hear of the first case of West Nile killing a crow in Maine. It will happen in my town, and I'll fret like a mother hen until I see all six of my crows together. This will take a tense two weeks. And I'll wonder if they're splitting up on purpose, to yank my tail feathers.

2

BUGS IN MY BELFRY

SHE'S AS WINSOME as a puppy! She's dexterous as a cat! She's cute as a bug! She is a bug!

The ladybug was hunting aphids on a bamboo leaf when I disrupted her with a tap. She turtled and rolled into my hand. I dropped her into a ceramic soufflé cup and slid her under the lens. Unfortunately, I had earlier put a slug in this same dish, and I didn't rinse it well enough. Now one of the ladybug's legs has been glopped with what looks like beaten egg. She lets me admire her undercarriage for a few minutes. Her black legs fold across her abdomen like armor, fitting inside the rim of her red wing-covers. Then she unclasps a couple of appendages and paws the air for traction. I give her a stick, and she scrambles upright. She rests for a good five minutes. One blunt antenna twitches. A front leg spasms. I'm afraid she may be processing a stress chemical that swamped her when I caught her. After a few more minutes, she pulls it together. She lifts a front leg and rotates her black-and-white face toward it. To my surprise, she bites onto the upper leg and pulls the appendage through her maw, cleaning it. She rubs the leg over her head exactly as a cat would, then cleans it again in her mouth parts. She cleans the other front leg. The rear leg is now glued to her red shell with dry slug slime. She works it free, then scrapes it against the edge of her shell (a pair of specialized forewings, really) until the yellow stuff flakes off. Grooming complete, she marches up the stick, hoists her forewings, unfolds her flying wings, and blasts off. She bonks into the microscope lens and falls on the deck. On her second try she whirs into the air and heads back to work.

31

A slug, for the record, is also stunning under the microscope, a glistening caramel mound with a faintly striped mantle. Even the brown bullets that encase mosquito larvae are beautiful, viewed large. And though some of the creepy-crawlies I slide under the microscope make me want to yelp and run away, I persevere in collecting my backyard bugs. I'm working up to a visit from the Maine Entomological Society. They're professionals. They're going to strain, sift, even suck, insects from all corners of the yard. I don't want to do any squealing when they're here.

I heard about the society from Chuck Lubelczyk, who has begun trapping my mosquitoes to test them for West Nile virus. A rugged, blond biologist with Maine Medical Center's Vector-borne Disease Lab, Chuck turned his attention to mosquitoes after the sudden arrival of West Nile. Last summer, seven dead birds in Maine tested positive for the virus. This year, no one expects good news. I've given up the pretense that my interest in crows is scientific. I know each member of the family that shares my yard. They're loud, they're unruly, and they're mean to other birds. But they've become pets, of a sort. My yard would be too quiet if they died.

"Initially people didn't think West Nile was going to get up here. They thought Maine's too cold," Chuck says as he hangs a trap in my old crab apple tree one evening in early May. A sterling lizard is bolted to his earlobe. "It'll be interesting to see how far north it can go. It's south as far as Florida now. It could get west to the Mississippi this year." In hindsight, this popular estimate misjudged the virus's mobility by half a continent. By autumn, crows will be falling into backyards from here to the Rockies. Well over a hundred other species of bird, including hummingbirds, owls, eagles, herons, wild turkeys, and ducks, will test positive, too. And by the time insect season withers, 284 people will be dead.

Chuck's mosquito trap lures female mosquitoes with a combination of light and carbon dioxide. Then a four-inch fan whisks them into a mesh bag. "The fan is timed to let something as small as a mosquito get past. But moths . . ." Chuck grimaces. "Moth purée. Or, they get through, and they thrash around and mess up the mosquitoes." Not every species of mosquito transmits West Nile, so the lab saves time by weeding out innocent species. But the identification of mosquito species sometimes comes down to counting tiny body scales and hairs under the microscope. And if a moth has spattered its own scales, or messier body parts, on the skeeters, the night's trapping may be wasted. Before he leaves, Chuck sets an egg trap. This is a plastic tub of smelly water. One of the West Nile mosquitoes, *Culex pipiens,* prefers a putrid pool for its progeny—the sort of brew you might find in a stagnant birdbath or forgotten kiddie pool.

The next morning when he collects the traps, Chuck harvests both adults and eggs. Standing together like tiny missiles, a hundred eggs make up a black egg raft on the water. Left alone, each missile would release a larva into the pool. The fat-headed larvae would grow large eating microorganisms and tiny animals—including each other, should options run short. If you forget to feed a batch of mosquito larvae in the lab, Chuck muses, kneeling to dip the egg rafts into a vial, you end up with one large larva.

A state lab will test my mosquitoes and eggs for virus. A diseased mosquito will bode ill for people, worse for crows. It will also darken my opinion of English sparrows. The sparrows share an ancestral home with West Nile virus, so they have some immunity to it. Unlike crows, who sicken and die quickly, a West Nile–infected English sparrow sits around with lots of virus in its blood for five days. Then its immune system may defeat the bug. During those five days, any mosquito that gnaws on the sparrow can tank up on the virus. And an infected mosquito can stay infectious even as it overwinters in my attic.

If a mosquito were to inject me with West Nile, I probably wouldn't notice. Fewer than one percent of people who get infected become seriously ill. The other ninety-nine percent gain lifelong immunity, so I offer my arm to all passing mosquitoes. Crows are another story. In two different studies, every crow that was purposely infected with the virus died. In one of those studies, even uninfected crows in the same cage died of West Nile, having

33

caught it directly from the purposely infected birds. This year as the plague flutters and flaps westward, a sickened crow will fall on the White House lawn. Anecdotal reports from scientists will echo down the Eastern seaboard: Crow numbers are crashing. A U.S. senator will wonder aloud whether the outbreak is the work of terrorists. Wildlife biologists will continue to gather new species of felled birds from backyards and deep woods. In Ohio, unofficial reports will suggest that upward of a hundred owls died in one week. Raptors at nature centers and zoos will perish. And every time I can count six crows in the trees around my yard, I'll thank the mosquitoes for sparing them.

It's deadly dry August by the time Chuck and some fellow bug-hunters from the Maine Entomological Society congregate in my yard. The first question I ask the president and state entomologist, Dick Dearborn, is how my yard differs from a natural area. He squints at the lawn through his glasses. His T-shirt is crossed by the band of an army-surplus collecting bag, and by the suspenders that hold up his green work pants. Sunlight glares off his head.

"I don't find this unnatural," he counters. "You've got a variety of habitats in a limited range. You've got pollen sources, nectar sources. You've got everything from wet to dry, vegetated to not-vegetated. That lawn, to an insect, is a grassland. Your trees are a forest. I don't think they notice the difference between urban and rural settings."

Although I haven't thought of it this way, I have discovered that different insects travel different parts of the yard. On one of his mosquito-trapping visits, Chuck had told me how to make pitfall traps, promising they would reveal a hidden world of creepy-crawlies. Until then, the state of the art for my insect-study program had been to lie on a chaise lounge and stare at the ground. This amused the neighbors. "Hey," I heard Neighbor Hugh stage-marvel from his side of the lilac hedge one afternoon. "Here's an orange kankamangus bug!" I stifled a giggle. "Well!" he yelled in mock-amazement. "An Ethiopian mite!" Also from my chaise time I had realized that, from the perspective of a crawling insect, my yard must constitute a vast nation. For many of them, crossing from one side to the other must require a Lewis-and-Clark-size ambition.

One day I watched the struggles of a single brown ant. The ant had a tiny prize clenched in her mandibles, and this greatly complicated her journey. (Like their relatives the bees and the wasps, ants are predominantly female. Males are produced rarely, and for just one purpose. They typically have big eyes for locating virgin queens, and fast wings for catching them.) As the ant attempted to squeeze between two blades of grass, her prize jammed. She backed and tried again. Then she reversed and retraced her path for about three minutes, covering two inches. She turned and tried another route. To her, the grass blades were as big as fallen trees tangled by a tornado. She struggled over and around and through them. In twenty minutes she progressed six inches. I tried to clear a path for her with a twig, but this frightened her into dropping the food. She spent five minutes hunting for it. A rice-size black beetle trundled by her in the fallen forest. Two large black ants passed. A gnat a fraction of my ant's size ascended a blade of grass, paused, and then spread its wings. A wee wind blew the gnat four inches sideways, where it crashed into the grass. It climbed another blade and blew another four inches.

This is an enjoyable way to spend summer hours. However, many teeny beings are nocturnal, and I found night work to be chilly and fraught with skunks. So at Chuck's behest, I ran a trapline. I dug four holes, in different states of my backyard nation: one in the shady state of Dead Leaves; one in Shady Grass; one in Sunny Grass; and one in Sunny Garden. Into each hole I plunked a plastic cup holding an inch of rubbing alcohol. Sorry, bugs.

The next morning I emptied the Sunny Grass cup onto a paper towel and slid it under my microscope. It was mobbed with ants— nineteen of them. There were also an earwig, three small spiders, a black beetle, and a tiny fly. The beetle appeared to have crustaceans (probably harmless hitchhikers, Dick Dearborn told me later) glued to its wing covers. I shifted the paper towel, scanning clods of dirt that had tumbled in with my subjects. The ants' eyes were enormous and almond-shaped. The earwig's pincers looked ferocious. A gust of wind lifted the paper towel toward the lens of the microscope, and as pincers and hairs and bulging eyes came at me through the eyepiece, I squealed and fell off the back of my stool. (Unfortunately, this would become something of a theme, and I would do it most often in the presence of impressionable young girls. After showing one girl around the yard, I picked up a rice cake from the ground and, spying twenty ants on its flip side, shrieked and flung the rice cake. After showing another girl how sparkly and gooey flowers are under the microscope, I picked up a dead black beetle to show her. On my palm, the beetle vibrated. I yelled and pitched it away. My squealy tendencies have been hard to dampen.)

My next pitfall trap, the Garden, introduced me to that Bane of Pitfall Trappers, the slug. Two leopard-spotted, inch-long specimens had fallen in and turned to latex. Everything else that fell in had stuck to them. I could discern one sow bug. I slung the whole mess into the bushes. Shady Grass was also sluggified, though not as badly. When I slid this batch under the scope, I discovered a herd of baby sow bugs, nearly cute in their pale teensyness. But Dead Leaves was best. A half-inch beetle with a metallic-green head and shoulders shone like a Christmas ornament. Butterflies don't come any prettier. There were also two earwigs, and a small, red-eyed surprise: "Fruit flies," I wrote in my notes, "are wild animals!"

As the days passed, the trapline revealed a pattern. The sunniest cups reaped drought-hardy ants and spiders. Sow bugs and slugs I caught only in the shady sites. During the day, slugs cower under dead leaves and detritus to stay moist. Sometimes I would see them at dawn, finishing a meal of dandelion bud, or sliding toward the Bamboo Wilderness. As for the sow bugs, I learned that they evolved from sea creatures and, although they have come ashore, haven't yet evolved an alternative to gills. To breathe, they have to stay damp. Beetles also turned up in the shady states, perhaps because that's where the food is, or maybe because they're poor at shedding heat.

Now, with the society here to comb my yard, the drought has become severe. Dick kneels in Shady Grass and unfolds a yellow pillowcase. He pulls a tulip-bulb planter from his bag and grinds it into the soil. He dumps a plug of soil on the pillowcase.

"Very dry." He spreads the powder, looking for motion. "Very, very dry. There should be root weevils, thrips, chinch bugs." He tears roots from the plug. Nothing moves. He brushes an acorn from under his elbow. "You've probably got acorn weevils," he says hopefully. Across the lawn we go, digging plugs, but the result is the same. "Not even any thrips on the clover. And the worms must be down to China." Chuck Peters, a high school science teacher, joins us, holding out a shiny brown fragment. "This is a piece of a click bug," he says.

"Is this a root-weevil wing?" someone else asks.

Chuck Peters drops his beetle bit. "You know we're desperate when we're looking at parts." But as they fan out across the Insect Nation, the bug hunters discover that not every state is equally depopulated by drought.

"Look at all that life," Chuck Peters breathes, lifting the lid of the compost bin. The oasis of my vegetable scraps is churning with sow bugs. A little later he hails me from behind the raspberries. He pushes aside a wad of straw mulch and starts grabbing at things. "This microhabitat is just teeming with life!" He pinches up a narrow rove beetle and tucks it into a vial to preserve it. Rove beetles bite, but a bug hunter doesn't care. A centipede rumbles across the exposed earth and dives into the mulch. Chuck pokes it. They bite, too.

"Bagworms like your shed," Dick says cryptically, standing by my miniature metal barn. I follow his finger and find a half-inch tube made from bits of leaf. It's stuck at one end to the steel wall. I notice a dozen more. Like caddis flies that make underwater homes of grit, these insects haul around houses of organic junk, eventually parking to emerge as mini-moths, Dick says. (I'm kind of embarrassed to admit I've occasionally found them on my walls indoors and wondered what they were. I have an open-door policy in the summer, and this year everything from bumblebees to a chipmunk will amble through my house.) Chuck Lubelczyk is hunched inside the shed, delighting in spiders. One lady's eggs have just hatched, and an astronomical number of specks are spreading from the papery sack. Small wasp nests hang in the corners.

We all converge at the raspberry mulch in the heat of noon. Dick sprawls on his stomach, peels back some mulch, and plows a furrow with his finger. Things appear. "Rove beetle," Chuck Peters says, handing it to Dick, who's collecting them for research. Something green lands on Dick's hand. "There's a neat bug," he says, and Chuck Peters raises his eyebrows at me. "When Dick says bug, he means bug. Order Hemiptera." Many insects have two sets of wings; in true bugs, part or all of the forewings are hardened into a flattish, oval shell. Bugs also suck their food through four straws that constitute a mouth. Most drill into plants, but the bedbug is a human specialist. Mosquitoes, by contrast, are flies. A copper-shiny beetle bashes across the dirt like a dinosaur. Dick tells me it eats earthworms.

"This is the Huck Finn approach to collecting," Chuck Peters observes. "You find a place where you're happy and you flip over a bit of stuff and watch things scurry across."

"And it looks like so much fun," Chuck Lubelczyk adds, "that people come and join you."

Huck and company move on, to examine an apple tree. "Carpenter ant," Dick says, pointing. His glasses are sliding in sweat. "Spider. Psocid." The last is a tiny mothlike guy sitting motionless in the topography of bark. "They feed on fungi."

Chuck Peters sees a Stealth-bomber-shaped fly he wants for his personal collection and deploys his aspirator. This is a stoppered bottle with two tubes going through the cork. He aims one tube at the fly and inhales sharply on the other tube. The fly lands in the bottle. A screen over the sucking tube prevents the fly from proceeding to Chuck's mouth. Chuck's students delight in getting their hands on his aspirator and secretly removing the screen.

Eventually, the sun becomes unbearable even for animals who cool their surfaces with sweat. Chuck Peters folds his equipment. He has strung a tarp like a sail between two bushes. While we bug-hunted, a dozen flies, beetles, and bugs have flown into it, rolled down, and splashed into a plastic half-pipe filled with soapy water. Booty. Chuck has also taken a kitelike span of nylon and gone beating the bushes with a stick. Bounteous bugs tumbled out of hiding and landed on the kite. This, too, folds to go in his car. The bug hunters then buzz away. The bugs and I are left with the sun.

Late summer is always a tough time for creepy-crawlies. Although they're solidly classified as cold-blooded animals, many of them do generate their own heat. Bumblebees shiver to heat up their flying muscles on a cold morning. Dragonflies shiver as they perch, to stay revved and ready to whir after a passing mosquito. Staying cool is a problem, especially for flying insects whose muscles produce waste heat. So midsummer is hard on bugs. On people, too. Now that I know how to look for bugs, I'll undertake a more methodical survey of my Insect Nation—on a cooler day.

On a gentler morning I'm circling the yard when I catch a whiff of nutmeg and follow it to a small hedge of bamboo (aka Japanese knotweed) on my north fence. Sprays of the tiny, white flowers are opening on the hedge. Like other plants that rely on animals to pollinate them, bamboo releases perfume to attract attention. The bamboo may also turn up the temperature of its tissues, a newly discovered ability among plants, which may help to lure insects. A savvy insect knows that a reward—nectar, nutritious pollen, or a free warm-up—awaits anyone willing to transfer a few grains of pollen from one blossom to another.

Looking at the bamboo today, you wouldn't guess that there's a pollinator crisis upon the land. But it's a fact. Some wild plants aren't able to recruit the assistance they need and go unfertilized. The cause is largely chemical: Pesticides that soak modern lawns, gardens, and farms aren't all discriminating about what they kill.

Bees, bugs, flies, and butterflies die along with the Japanese beetles and the cutworms. Since some plants have evolved to be picky about their pollinator, the loss of a particular insect can frustrate a plant's efforts to reproduce. My favorite example of this problem unfolded when French colonists took Mexican vanilla plants to Réunion, off the coast of Africa. No native insect would deign to visit the new orchid, however, and the Mexican bee who provided the service at home refused to travel. So all vanilla grown outside of Central America (and most of it grown *in* Central America, for that matter) is pollinated by people, by hand, one flower at a time.

The bamboo is having no such trouble, though it's not native to North America. Its spicy summons has gone out, and a blur of insects has assembled. On my side of the hedge I count a dozen honeybees and nearly twice that many bumblebees. White-faced wasps, and an orange-and-black wasp with a glittering gold face, all paw through the flowers. A yellow jacket in a nectar frenzy lands on my wrist and bites me—doesn't sting as a rational yellow jacket would, but bites as a nectar-crazed one evidently will. Flies of all shapes and sizes work the blossoms. Whether they're harvesting nectar, pollen, or other insects, I can't tell. I pick a few of the flowers and take them to the microscope on the deck. I can't see any pollen, but the flowers are slick with nectar. And there are, by my quick calculation, more than two million individual blooms in this small hedge. The Bamboo Wilderness on the other side of the yard must hold hundreds of millions more. That's a lot of calories.

The bamboo festival lasts for days. The bees are so intent on harvesting nectar that I can pat them as they work. I can ruffle the yellow fur on a bumble's thorax. I can stroke the black corduroy of her abdomen. The honeys are busier and more intimidating, but they're equally pattable. That busyness of the honeybee is part of the pollination problem. When Europeans brought hives of these sugar-makers to the Americas, the honeys made a beeline for the niche occupied by native bumblebees. They visited the same flowers, they gathered the same nectar, and they did it more efficiently. Honeybees excel at finding thick concentrations of the sweetest flowers. And thanks in part to the bee boogie they perform back in the hive, they excel at leading masses of their sisters back to suck a flower patch dry. As a result, wherever honeybees roam, bumblebees can't afford to raise as many new queens as they used to. In his classic book, *Bumblebee Economics,* zoologist Bernd Heinrich calculates an unseen cost for the pint of honey in my cupboard: Had the nectar that went into that honey been collected by native bumblebees instead of European honeybees, it would have supported 436 more "reproductives," the young males and queenlets who go forth in autumn to start new bumble families.

Just as native bees are losing nectar, native plants are losing native bees. In a study of the pink lady's slipper, an orchid that grows in the woods around here, scientists found that only two percent of the flowers in the study lured a bumblebee and achieved pollination. The fertility of more than half the plants subjected to similar studies has proven to be limited not by poor soil or marauding caterpillars, but by a shortage of pollinators. If honeybees have elbowed aside the bumblebees, why can't those honeybees step in and provide the pollination service? One reason is related to the vanilla story: Some complicated flowers can be penetrated only by a specialized insect or bird. If that specialist becomes rare, the flower is out of luck. The other reason is that, after the imported honeys hammered the native pollinator population, their own population crashed. Two plagues of bloodsucking mites slew huge numbers of honeybee colonies, perhaps including nearly all the feral families. Beekeepers have lost most of their domestic colonies, too. If the native pollinators can overcome pesticides and the paving of their habitat (bumble queens raise their families underground), perhaps now is their chance to reclaim their niche.

In my bamboo hedge, though, the pollination system grows more complex by the day. One day, two scarlet dragonflies arrive to perch in the bamboo and wait for the right-size meal to wing by. The dragonflies are as spleeny as hummingbirds and spend more time arguing than hunting. A day later, a little gray bird, a least flycatcher, takes up residence in the nearby apple tree. He swoops into the humming airspace to snatch an insect, then returns to his branch.

When the bamboo flowers turn brown after ten days, the birds and dragonflies will have been fed, the parasitic wasps and flies may have slapped an egg or two onto a host, and the bamboo will be pollinated. Those insects who have harvested sufficient nectar and pollen will plow their resources into a new generation. Underground, maybe even under my deck, a bumblebee will top up her mud honey-pots and fatten her brood.

The next stop on my survey provides a lesson in insect sex. Although insects are wild animals, they're not shy about using the microhabitats right against my house, even for their most intimate activities. One afternoon I hear a cricket gritching beside the deck. I steal toward the sound, expecting it to stop. For once, it doesn't. This is a cricket with something even more important than safety on his mind. This is a cricket in love. I sit slowly on the steps and watch a cricket version of the Trojan War rage at my feet.

Paris, I presume, is standing atop one of the rocks lining the garden. He flicks his right wing over his left, scraping the tines of a little comb over a little bar. *Crick! Crick! Crick!* A whole herd of shiny, black Helens, with tubular ovipositors at their posterior ends, muddle around in the garden, eating fallen tomatoes and other detritus.

Across the plains of the garden comes Menelaus. He hardly slows as he passes the Helens. At the base of Paris' rock he issues a mighty *Crick!* Paris tumbles down to face him. The warriors glare at each other. And then they lurch forward, propelled on their mighty legs. Their black heads collide and they grapple with their forelegs. Each tries to heave the other backward. Even as they heave, they *crick*. Again and again they butt and heave, until Menelaus backs away and quits the field. Paris will hold Troy for another day.

He surveys the spoils of war. Doing a bucking dance, he approaches a Helen. She is intent on an old tomato. Another Helen burrows under a scrap of clear plastic that has settled near the chives. He *cricks* behind a third lady, then taps her wings with his antennae. She doesn't leave, so he walks onto her back. But her ovipositor is buried in the dirt and she's laying eggs. He walks off over her head. From under the deck another male *cricks* at half-volume. I wonder if he's trying to attract a girl while reducing the odds of getting his cricket teeth kicked in by Paris. Paris finally returns to his rock, followed by a Helen. After some coyness at the entrance to a tiny grass-cave beside the rock, she enters, and he hops down after her. They reemerge shortly, and the damsel returns to the garden.

Across the plains comes another challenger. Crickets rarely hop, I notice. They plod along like the rest of us. The new warrior walks straight to Paris at the cave entrance, but turns tail when Paris comes out bucking. I pull this guy out of the grass and set him on my knee. He's a handsome devil, with shiny black eyes and long, sensitive antennae. He's missing one of his jumping legs, which might explain his aversion to conflict. He sits in the sun for five minutes, then hops down. Five males now mill among the females, too many for Paris to fight. But most of the Helens are already laying eggs anyway, loosening the dirt with their forelegs, and poking their ovipositors into the soil. All of these heroes will die with the first hard frost. Then, if I do no destructive gardening, micro-crickets will emerge in the spring. They'll molt as many as a dozen times before they're mature. And a new generation will sport on the Plains of Troy.

The cricket's approach to sex is so straightforward as to be aberrant in the insect world. At the other end of the spectrum are the springtails. Springtails are mini-insects who resemble, if anything, elongated fleas. But they're tiny. Nearly seven hundred species inhabit North America, but most people know them only as spots of dirt that move. According to my rough estimate, I could be stepping on a few hundred each time I set a foot on the lawn. In addition to the usual six legs, many springtails feature a tail that can snap them into the air. They also feature notable mating methods. In one species, the male approaches his much larger lady and grips her by the antennae. He deposits a droplet of sperm on the ground, then drags his date onto it. In another species, residents of trees, the male deposits a corral of sperm droplets, each upon a stalk, around his quarry. When she exits the corral, she is bound to collide with a droplet, inseminating herself. And these are the more romantic species of springtails. Most of the males leave their sperm stalks any old place, for any old girl to stumble across. Scientists have watched males tend these sperm gardens, in which they appear to eat their stale deposits and plant fresh ones.

Studying the sex lives of insects, I finally decipher the mystery of looped dragonflies. Many times I've witnessed two dragonflies locked together in a loop, his tail grasping her head, and her tail curled under herself and locked onto his chest. Yes, they're mating. At some point in his evolution, the male dragonfly decided that the wisest thing to do with his sperm was to bend his tail forward and deposit it in a reservoir on his chest. Now to collect it, the female has to maneuver her own tail up to his chest. In some cases it takes an hour to complete the transfer. How this sort of thing evolves, what series of genetic mutations culminates in such an arrangement, is one of the ten million questions that keep evolutionary biologists awake at night.

✺

But back to my survey. The creepy-crawly habitat comes closer still to my own domicile. A spider has decided her ideal microhabitat is my office window, across which she has stretched a web. Eek! All those grabby legs! That hunching abdomen! I bear no ill will toward spiders, but they do give me an urge to screech. Two panes of glass separate us, and that helps. I glance at her. She's head down, in the center of her web. Her eight legs radiate around her. She sways with the cool wind.

I raise a hand to my side of the window. She's not delighted to make my acquaintance. Like a washing machine starting its spin cycle, she bounces in the center of her web until she's whirling in a circle. If I were a bird, would I now think her too large to eat? Or too ferocious? Unlike me, she conquers her phobia in a single encounter. The next day I open my hand at the glass again, and she's unmoved. I breathe deep and examine her in detail. Her abdomen, the size of a pea, appears soft with brown fur. A trail of white dots runs down the middle of her back, with a line of chocolate dots on each side. Her legs are checkered brown, black, and white. When she turns her underside to me, her usual position, I can see the thinnest silver rope connecting her spinnerets to the center of the web. She is an elegant creature. To soothe my psyche I give her a round and pleasing name: Babbette.

The web itself is in the classic orb-weaver style. I count twenty-nine spokes, and each spoke is crossed by twenty-six rungs. The silk itself looks a bit fuzzy. Nose to the glass, I discover that the rungs are all strung with minuscule pearls, nearly beyond the limit of sight. I remember reading about this: One of the six or eight silks that orb weavers can produce forms sticky beads. When it comes out of the spider, it's coated with salts that attract moisture from the air. As water gathers on the silk, gluey beads form. The wet beads trap insects and also keep the web taut, yet stretchy. The spokes themselves are made from a different silk and aren't sticky.

When a fly collides with the web, Babbette grips two spokes of her web, clamping them in tiny claws. She tugs as if testing a fishing line. She likes what she feels. She trundles down the spokes and grabs the fly. With her rear legs she pulls silk from the nozzles under her abdomen and whips it around the fly. Then she sinks in her fangs to inject a pulse of poison. Carrying the wrapped meat in her mouth, she returns to the center of the web. She alternately spits chemicals into the fly and sucks out fly soup. In the web, gaping holes mark her path. Each time her foot fell on a gluey rung, she destroyed it. And the next morning the web is gone completely. I'm a tiny bit disappointed.

A day later, though, the web is back. This establishes a pattern. Whenever Babbette has caught a few meals and trashed her web, she cleans it out. She leaves the thick ropes that frame the orb and moor it to my house. Then she applies new spokes and lays on the sticky rungs. I catch her at this only once and watch her pull two strands of silk from her spinnerets with both back legs.

45

The two strands merge, and when she's produced enough, she dabs her abdomen against a spoke, fastening a rung. Over an hour, she repeats this 750 times. The freshest rungs are smooth and silvery. But by the time she's halfway to the center, the first-laid rungs are pearling with glue. I never catch her clearing out an old web. She does it in the wee hours of the night, evidently, or when she knows I'm out of town. Many spiders eat the silk when they remake their webs, recycling their extraordinary proteins. Babbette is too cagey to reveal her secrets.

She's not shy about eating insects, though. She eats a lot of them. One day she scores a large robber fly. This inch-long fellow won my admiration after I read about the trials he endures when mating: The females are such knee-jerk assassins that a courting male either holds out food or waits until the lady already has a mouthful of something else. I suppose it's a spider-eat-fly-eat-gnat world, but I was sorry to see Babbette straitjacket the big boy, suck him empty, and cast his remains onto the heap of exoskeletons that must be accruing on the cellar door below. Far more disturbing was the housefly who spewed maggots. As Babbette tucked into this meal, I noticed something white wriggling out of the fly's abdomen. Huh? Out came a maggot, waving its tiny head, as though it had never dreamed of hatching twenty feet in the air on a sticky trampoline. It was followed by another. And another. By the time this horror movie was over, the fly had disgorged seven larvae, probably the progeny of a parasitic wasp or fly. All plummeted to the carcass pile, leaving Babbette's feast diminished.

My spider's behavior will become erratic toward the end of the summer. She'll sleep late on cold mornings, emerging from the eaves at noon. Sometimes she'll leave her web unmended for a day or two, perhaps to molt. Her sisters, whom I'll discover nearby, will do much the same. Below Babbette in the garden, one sister will spend days on end disguised as part of a dead plant stalk, minimizing her odds of being spotted by a bird. And on the front porch, a third sister will retreat to the ceiling, scuttling down to her web only when she senses a good, juicy reason. Each will swell to the size of a peanut.

Babbette's a she-spider, saving up calories to convert to eggs. Male spiders are usually small, and they quit the web-making business as soon as they make a final molt into maturity. They turn their energies to mating. A male will load his sperm into his palps—antenna-like wands under his chin—then approach a likely web. To avoid being mistaken for the blue plate special, he'll pluck the web with a species-specific rhythm. He might build a special mating thread onto the web and invite the lady onto it. Then she might invite him onto her web, where, after long, spidery conversations, he might be permitted to empty his palps into a pouch on her underside. A few weeks after Babbette mates, she'll produce about a thousand eggs and wrap them in the appropriate type of silk. And then, unless she's one of the few overwintering species, she'll die.

In September, Babbette will disappear from my window. My only souvenir will be a remnant of web that a storm plasters to the glass. For a while, the outline of the strands will remain, with a film of glue oozing down the pane. Then more storms will pelt the window with snow, and the last of Babbette's web will vanish. Maybe next year her spiderlings will emerge from the eaves to continue my desensitization.

My tour of habitats confirms that if there is one truly global citizen of the insect world, it is the ant. I find ants high in the trees, deep in the ground, all through the grass, and under every rock I turn over. They farm aphids on the tomato plants, gathering honeydew excretions in payment for bodyguard services. They collect a nectar fec from tiny nectaries on the peony buds, again in exchange for fighting off harmful insects. They carry daisy seeds across the walkway and into the ground. Ants are ubiquitous. I had asked Dick Dearborn if the ants in my yard collectively outweigh me. "Probably."

One day I find a train of ants climbing the pole to the hummingbird feeder. I pull up a chair to watch. A few are large and black, but most are a zippy, golden variety. Trafficking up and down the pole, they stop and tap antennae with every other ant they meet. Those who are ascending make their way onto the top of the feeder, then crawl down the plastic globe, their feet searching for a grip. They negotiate the overhang where the bottle narrows into a red disk. And there, where a film of sugar water glimmers between bottle and disk, they wedge themselves to drink. The base of each yellow flower is also ringed with drinking ants. And a stream of ants, abdomens glowing pink with sugar water, returns up the bottle and down the pole. Back in Zippytown, they'll cough up the goods, just as bumblebees and honeybees do, for their kin to consume. That they're feeding here, four feet off the ground, means that some enterprising scout trundled all the way up here by herself, just checking, just making sure no food was hanging around uncollected. When she found it, she laid a chemical trail back to Zippytown, and the working caste hopped on the trail.

When I sit at the base of the pole to locate Zippytown, I'm distracted by the plight of a gnat. It landed on the pole and somehow became stuck there. It can't, or won't, fly away. Every ant that ascends the pole frisks this gnat before continuing. Finally, an ant walks over the gnat's back, patting and prodding, and grabs a leg. When the ant tugs, the gnat slides. Encouraged, the ant hugs the gnat's abdomen and hoists it clear of the pole. The gnat, wings unfurled in the ant's face, makes an uncooperative load, but slowly the ant turns to descend the pole. Puffs of wind toss the gnat's wings. The ant's feet shiver over the metal, wanting traction. One after another those feet release. Ant and gnat blow off the pole and into the garden. From what I've seen of ant work, it could take the ant the rest of the afternoon to get this meal home.

What species these are, I couldn't say. North America is home to about six hundred kinds of ant. Most eat a varied diet, taking gnats and sugar water where they find it. Other species consume only the eggs of spiders and other arthropods. Or only seeds. Or only other ants. A few farming species consume only fungus that grows on the leaves they shred, or the honeydew excreted by aphids they tend like herds of cattle. Few are helpful pollinators; more are nectar stealers. Most are Nature's little tillers, aerating and mixing the soil.

The black ants on the pole are mellower than the Zippytown ants. Perhaps their diet runs more to seeds than sugars. To test the theory, I load a chopstick with crystallized honey and dab blobs on a rock at the base of the pole. In a few minutes two Zippytowners and two Mellowburgers find the blobs. But after fifteen minutes Zippytown's honey is a scene of frenzy, while just four Mellowburgers mill around their blob. And after thirty minutes, Mellowburg has rolled up the sidewalks and all the ants have gone home. Except one. She's claimed the chopstick itself. I carry her to the microscope.

Magnified, she's not black, but mahogany. Her large eye is grainy with individual lenses. Her face ends in something of a beak. This is pushed into a cavity she has mined in the honey. Beneath the beak, all hell is breaking loose. In contrast to a human mouth, the ant's mouth is crammed with tools. Half a dozen palps and jaw parts seem to be involved, some flailing at the honey, others shoveling it in. Liquefied honey floods her open mouth. Every few seconds she pauses to swallow. As if telegraphing her bliss, one antenna and one foot drum on the chopstick. I set her back on the rocks of Mellowburg.

On another day, three winged ants struggle up the lawn chair I'm sitting in and take to the sky. It's mating day. Most species of ants send forth their winged reproductives on the same day. In some species, the males of many different anthills will swarm in one place—above a tree, or over a roof, for instance. They'll slap on some pheromone aftershave, and the girls will come flying. In other species, the girls exit the nest, fly to a homey spot, and then they do the perfuming. Either way, the result tends to be a lot of winged ants in the air at a time. The vast majority will be eaten by birds, reptiles, or even ants and other insects. Or they'll drown, or fail to find a mate, or I'll set a chair on them. But that's why there are so many. One is bound to get lucky. A mated queen will shed her wings, dig her burrow, and start a family.

❦

Lately when I step onto the lawn, I think about how many little beetles are holding their breath, pressed between my foot and the baked soil. How many spiders flatten in the grass, waiting for me to move on? How many sow bugs? How many ants drop their burden, then spend an hour hunting for it? Although I do worry about squashing them, I'm happy to learn that so many insects are at home in my backyard ecosystem.

"Insects are versatile," Dick Dearborn tells me. "I don't think they notice the difference between urban and rural settings. They've got a lot of different things to feed on, a lot of places to hide. Almost every species of mosquito in the state is here. And I'd say the number of ground-beetle species is about the same as anywhere else, too."

At least that's the case in my yard, where no pesticides soak the greenery. On some lawns this summer, the usual toxic chemicals will be augmented by mosquito poisons, as some cities try to fight West Nile by fogging every living thing. The spraying campaigns are reigniting a decades-old fight between those who yearn to destroy one pest and those who want to protect thousands of innocent species. Dick expresses no ambivalence.

"Pesticides disrupt everything," he growls. "They're disastrous."

In Maine, West Nile flutters on unhampered by mosquito poison. In a ten-mile circle around my yard, mosquitoes infect crow after crow, and the birds, hemorrhaging, tumble to the ground. The red dots on my dead-crow map cluster close to my neighborhood. Perhaps mosquitoes in the deep woods are infecting as many birds, but no one is picking them up. Perhaps in this instance, suburbanites are on the front lines of observing a plague as it sweeps through the natural world.

50

3

LAWN OF MANY WATERS

THIS YEAR, APRIL showers didn't promise May flowers. The ground didn't thaw until the end of March, so the melting snow ran down the street instead of soaking the soil. Then, when the ground was soft, the rain quit. Last year was droughty, too. Out in the country, wells are going dry as if water were going out of style.

Here in town, the drought strikes only one of the three water sources in my yard, the water that falls from the sky. A separate river system, stolen from a lake north of here, flows through the neighborhood inside iron pipes. And a third source lies untapped ninety-six feet under my turf—at least that's what the dowser will tell me.

The drought shouldn't be a disaster for my yard. Trees and shrubs can constrict the mouthlike stomates in their leaves to hold in moisture. They can put less water into their fruits, delaying reproduction for a year. They can grow less. My animals should be fine, too— and they could teach me a thing or two about water management strategies. As an animal, *Homo sapiens* relies heavily on water. I exhale a pint of it doing nothing more strenuous than sleeping at night. It exits my skin constantly. I pee extravagantly. To counteract my leakiness, I need about ten cups of liquid a day. I drink sixty percent of it and extract another thirty percent from my food. Metabolism in my cells cobbles together H's and O's to make the last ten percent. Other animals, especially small ones, and particularly desert dwellers, are more resourceful. The kangaroo rat can wring all his water from food and metabolism, and he retains that water by concentrating his urine five times more than my body can.

51

Even in temperate zones, some critters get much of their water from food and recycle that water carefully. The deer mice in my yard safeguard their moisture by working in the cool of night. They waste little water on their dry feces, and their pee is four times more concentrated than mine. My crows, instead of flushing urea waste from their bodies with a flood of water, concentrate it in their kidneys and eject a paste of white crystals along with their dark feces. Even squirrel pee is more concentrated than mine.

I know my animals evolved to survive drought, but I can't help helping them. After noticing the crows sitting in the shade with their bills gaping in the heat, I do something I've never bothered with before. I buy a birdbath. For the squirrels, mice, and ground birds, I also fill an old broiler pan and place it in the shade of the chokecherry tree. And when a young chipmunk becomes a regular visitor in my kitchen, I set a dish of water on the floor for him, too.

The chipmunk, the chipmunk. I catch sight of him first in the cold late-spring. He's gathering maple twirlies from the driveway beside the house and scooting under the front porch to stash them. I haven't seen a chipmunk here before. Having recently hand-fed one at my grandmother's apartment, I've been pining for one, though. And here he is. He's small and his tail is slim, suggesting he's a young lad dispersing out of his mother's territory. His copper fur, striped with black and white, is pristine, unflawed. He's in perpetual motion, a busy, busy boy.

I'd love to feed him, tame him, but I worry. It's not that I fear he'll become a pest. Chipmunks rarely invade walls, and I don't care if his tunnels disturb my haphazard flower beds. My concern is that I'll falsely influence his choice of territory. That I've never had a chipmunk before could mean that my yard doesn't meet their needs. There are definitely too many cats, and there may be other threats I'm unaware of. Maybe there's insufficient food or inadequate shelter. So I wait. If he chooses my yard on its own merits, then I'll mess with him.

He stays. So after two weeks I sit down on the deck with a handful of sunflower seeds. I toss the seeds loudly on the planks. Within five minutes the chipmunk pops up the steps and vacuums the seeds around me, bumping my hand in the process. With his cheeks bloated, he dashes into the lilac hedge to unload. I scatter more seeds and keep a few in my palm. When he returns, he cleans the deck, then sniffs around my hand a few times. He stretches the triangle of his head over my hand. I feel a slight pressure as his teeth and tongue gather a seed. Then another, and another. His front feet step onto my thumb. Then the back feet follow. He's perched in my palm, hoovering seeds. As he shifts them into his cheeks, the seeds make a dry, grinding sound. Bulging, he bolts. The next time he comes, I keep all the seeds in my palm. Up he hops to collect them. Slowly I move my thumb toward his head and stroke his fur. He leaps away, but comes back. I spend an hour with the chipmunk, raising and lowering my hand slowly with him aboard. His fur is silky velvet, nearly too soft to feel.

The next morning I surprise him in the kitchen. I habitually open the back door to let the day in, and he has availed himself of the invitation. "You're a cheeky bugger," I tell him as he skitters under the stove. I sit on the floor, seeds in hand, and wait. After a minute his deer-colored nose comes into view. I drop a seed loudly on the floor. The black eyes appear. I drop another. The ears are out. After a few false starts, Cheeky walks cautiously across the floor. He picks up a seed. When he locates my handful, he forgets his fear and gets busy. Then—zing—he's over my legs, out the door, and off the edge of the deck. I go about my business, only to catch a coppery blur zipping behind the refrigerator three minutes later. Cheeky's back, and his cheeks are empty.

Cheeky turns up the next morning, too, and the next, and the next. Day by day, I make him climb onto my lap to get his seeds, then onto my shoulder. Shaking a cup of seeds, I lead him to the top of a wooden stool where I leave the cup. Now when he arrives, he doesn't have to risk being stepped on to get my attention. I'll hear the *gritch-gritch* of seeds, and there he'll be, perched on the cup's rim, his body rolled into a stripy ball as he dives in.

Occasionally I'll confiscate the seeds and put him through his paces. During one session I count out 150 seeds to offer for each visit, to see how many he can shove into his face: He averages 77, and he's still a teenager. I try to vary his diet, offering banana chips, almonds, peanuts, and figs. He's finicky. If no seeds are available, he'll eat a few bites of fig, then haul the fruit away for storage. He'll sometimes sample a banana chip. The only serious competition to sunflower seeds comes from oranges. He'll grip a slice with his tiny front claws and tear at the juicy flesh. He smacks his lips, shreds of pulp hanging from his chin.

I soon habituate Cheeky to adoration. From thumb-stroking, I move to patting him with a finger. At first, he ducks away, although he doesn't stop gathering seeds. Then he gives in. I run a finger from his nose to his rump and let his bottle-brush tail slide through my hand. I raise him to my face. As he works at the seeds in my palm, I lower my nose to his fur. I give his round back a smooch. That I'm kissing a chipmunk makes me laugh, and Cheeky gives a little bounce of alarm. But he works on, weaseling the last seeds from the creases of my hand. I bend again and sniff his fur. He smells of earth and yarrow, a spicy weed that grows in my lawn. I wonder if he lines his bedroom with it, as starlings do.

Cheeky grows bolder. If no seeds wait on his stool in the morning, he goes into search mode. If I'm on the couch in the living room, he'll *pitta-pitta-pitta* in to find me, then pop up onto my knee. If my sweatshirt pocket holds seeds, he smells them and wriggles in. Like my crows, Cheeky becomes a friend. When I hear his feet *pitta-ing* through the house in the morning, I can call "Cheeky-Cheeks!" and he'll track me down. I can pat my thigh and he'll scramble up my leg and perch in my hand. Generally, he works for a couple of hours in the morning, then retires. If I let him, he can spirit away two cups of seeds. He drinks only rarely, and only about once a week do I notice a dime-size puddle on the floor. His water-saving poops are another matter. He's free and easy with those, especially when startled by a sudden noise. But they're dry and easy to manage. And I wouldn't care if they weren't. I love this chipmunk. There's something breathtaking about the trust of a wild animal. I'm flattered that this small creature can overlook the strangeness of my species and hang out with me.

The only hitch in our relationship is that seeds must be present. If Cheeky climbs my leg and finds no reward, he hurls himself to the floor with a disgusted splat. And he never makes eye contact. When he does look at me, it's to see if I'm wearing climbable clothing, or to measure his leap from my leg to my shoulder. Sometimes I wonder if, in his view, I'm just a strange kind of tree. But I don't care. Cheeky's my pal. And as the summer parches, I'm glad I can lend him a paw.

Even a dry spell here isn't rainless. And when the rain does fall, the whole yard seems to sigh in relief. Cardinals and crows shake their plumage in the sprinkles, catching a bath. The oak leaves shed a layer of dust that has settled on them. Down in the soil I imagine the springtails and earthworms are squirming happily in the moisture. The only malcontent is Cheeky. He won't work in the rain. He stays home. And even when the weather clears, some part of his tunnel complex must remain muddy.

The first time he pitters into the kitchen with red earth caked on his face, I laugh aloud and scare him under the stove. When he comes out and settles into his seed cup, I assess the damage. One ear is pasted tight against his head and filled with mud. Streaks of red mute his stripes. Gritty glue has hardened his cream-colored eyebrows into spikes. He looks like Evil Cheeky. He couldn't possibly care less. But I like my chippy soft and kissable. When he goes out to unload, I moisten a kitchen sponge and wait by the seeds.

"Squrrrr," Cheeky protests as I dab his ear. But he doesn't lift his head from the cup. I push harder, trying not to mash his face into the seeds. "Squrrrr!" The mud has set like cement. The sponge clogs. His ear is half-done when he packs in the final seed and sprints out the door. I rinse the sponge and resume on his next visit. It takes a good half hour, and a great deal of complaining, to clean him up. I can't get all the mud off his eyebrow, so he wears an enraged expression for the rest of the morning. The next day he's silky again. I expect the mud grooms off easily when it's dry. Animals who live in the elements must have evolved ways to deal with excess water, just as they have ways of dealing with shortage.

To combat scarcity, birds keep an eye peeled for the glint of water whenever they're cruising over the planet's surface. If they don't win a nesting territory that incorporates water, they have to commute to it. In my neighborhood they needn't commute far to find surface water. One spring day my ramblings lead me southwest a few blocks to a patch of woods. Surrounded by suburban homes, it's twenty acres of bedrock ridges and trees. Caught in the rocky troughs I find a series of shallow ponds. And flitting around the ponds are a series of warblers. These aren't residents, but migrants who have broken their journey to drink at these teacups.

Still, a commute of just a few blocks is risky for a little bird like a goldfinch or a chickadee. It exposes him to sharp-shinned hawks. In mating season, it leaves his wife unchaperoned. And later, it cuts into the time he could spend feeding his children. When I bought my birdbath, it became clear that water is a valuable commodity in the neighborhood. My little oasis was mobbed. Goldfinches, house finches, and cardinals gathered in the lilacs and the chokecherry tree, waiting a turn to drink. The catbirds bathed daily, flapping half the water out of the basin. And someone used the bath to soak their french fries. I had my suspicions, but I checked the Web site of crow scholar Kevin McGowan to confirm. Sure enough, the crows are using the public fountain as their Crock-Pot. From what I read, my crows are being fairly polite. Someone else wrote to McGowan's site about a backyard birdbath that's all a-stew with dead snakes and rodents. The crow man offers two possible explanations: One, when crows are feeding a nesting mom, they bring her water by soaking her vittles. Two, McGowan speculates, crows may just like their food a bit on the putrid side.

Between the splashy catbirds and the crows' bloated snacks, the birdbath wants freshening every day. The patrons of the broiler pan are more mannerly. Squirrels drink daintily from the edge. Blue jays bathe there, but less boisterously than catbirds. I dump the pan every few days in case nobody's eating the mosquito eggs that are deposited there at night.

As for my plant citizens, I give them not a drop of water. The oak tree, the black raspberries, and the native asters are surely adapted to waiting out a dry year. Everyone else can sink or swim. I'm not going to siphon water out of Sebago Lake to coddle English grass and clover.

Of the three waters flowing through my yard, surface water feels the drought quickest. The skin of the earth is drying. Less water is migrating through the soil. Streams, which gather half their volume from the ground around them, are shrinking. They're slow to refill the lakes. In suburbs and cities, the surface water is especially hard hit. Even in a wet year, the profusion of pavement sluices rainwater away from the soil. Rain that falls onto cement or asphalt usually flows into a gutter, then down a drain, and is piped away to the sewage-treatment plant. The soil is cheated of this moisture, and the loss is colossal. Scientists have calculated that pavement is costing the city of Dallas between 6 and 14 billion gallons of water a year, and Atlanta as much as 133 billion gallons. Underneath the asphalt umbrellas is parched soil, and a dwindling supply of groundwater.

Here in my gasping yard, dust and dirt are building up on both the paved driveway and the lawn. When rain does occasionally fall, perhaps seventy percent of it sinks into the sandy soil. If the day is hot, a fair amount evaporates. The remainder rolls down-hill, especially if it lands on slanted ground. This runoff washes squirrel dung and lawn-mower oil off my lawn, dust off the roof of my house, engine oil off my driveway, dog manure off the side-walk, grass fertilizer off one neighbor's yard, and dandelion killer off another's. The runoff also gathers heat from the roof, the tar,

57

and the ground. Hot and dirty water slips toward the ocean, following a fold in the landscape. When my runoff trickles into the beach sand and mingles with salt water, it bestows all its heat and grime upon the Atlantic. This is how *E. coli* bacteria, pesticides, fertilizers, and an *Exxon Valdez*'s worth of oil runs, spot by spot, quart by quart, into the water bodies of the United States each year.

As grubby as my yard sounds, I'm surprised to learn that farmland can also be a toxic tea bag. Lawn owners and farmers alike are prone to overdosing their crops with pesticides, herbicides, and fertilizers. If they scatter the chemicals before a rain, that much more will depart with the rolling water. Right now, the scariest ingredient in the tea bag is atrazine, a popular weed killer. Corn and sorghum farmers spray it in the spring, and rain carries it into the water soon thereafter. People drink it. Frogs grow up in it, and one scientist has shown that it can make them hermaphrodites. That's just one chemical. Nitrogen, in the forms of dung and fertilizer, also lies on the surface of farmland, waiting to be washed off. When too much nitrogen flows into a pond or a slow-moving stream, it overfeeds the resident algae, whose sub-sequent decomposition sucks oxygen from the water, suffocating fish and other creatures. So, while hot and overpaved towns like mine foul rainwater with heat, manure, and chemicals, they're not the only offender.

One solution to the gunk that washes off my yard is to capture the water and give it time to sink into the soil. The longer it spends in the company of soil microbes, the cleaner it will be. At the same time, it will rehydrate my soil, nourish my plants, dampen my insects, and keep my food chain chewing. I wade through a pile of research to see what the ideal driveway pavement might be, but conclude that almost anything that slows flowing water will work. I could dig a holding ditch on either side of my driveway. I could channel all the runoff into one swale. I could replace the asphalt with special pavers that let water sink between them. But I suppose the simplest plan might be the one I've already begun. I'm letting weeds colonize and destroy the existing pavement. As for the roof runoff, I suppose I ought to dig big holes under my downspouts and fill them with rocks and sand. Better yet, I'd channel the gutters into a cistern and use it to water the organic corn I'd grow where the lilacs are . . . Maybe next year.

In my yard, this water that falls from the sky is only one branch of a three-pronged river. It's the most visible third, and the third that most quickly reflects the character of my neighborhood. It's the third that keeps the top inches of the soil moist and fills low spots with water for squirrels, skunks, and birds to drink. It's the third that gathers the jetsam of town life and delivers it into the neighboring streams, lakes, and ocean. But it's only a third.

~

If I, like Cheeky and the crows, used water only for hydrating my food and flesh, I could probably get by on what falls from the sky. But I need more than the two quarts I drink each day. To shower, wash clothing, and flush the toilet, I used another fifty or sixty gallons a day. While this is half the amount used by the average American, it's still far more than my yard can supply.

To slake our cultural thirst, we human beings build our own metal-clad or masonry watersheds, which we direct into our towns. An artificial river now flows right through my house. This Cast Iron River starts fifteen miles northwest of my house at Sebago Lake. It splits into thousands of branches, one of which flows under my street and into my basement. It breaks into daylight from a number of faucets. Then, carrying my skin flakes and deodorant, my excreted caffeine and solid waste, the Cast Iron River plunges back down into the sewer. At the treatment plant downtown some of my contributions are removed. Then we've got no further use for this river, so we release it into Casco Bay.

Because I don't water my lawn, the Cast Iron River is almost completely separate from my natural watershed. I do transfer a few gallons a day from the Cast Iron to Nature, when I refill the bird-baths and Cheeky's dish. And I'm ambivalent about it. Rainwater isn't clean by any stretch—it rinses buckets of pollutants down

59

from the atmosphere. But tap water is even more intensely flavored. Traditionally, water agencies use chlorine to kill the bacteria that beavers, birds, and bathers deposit in surface water. Chlorine works brilliantly against germs, but it also reacts with the corpses of leaves and other organic dross in the water, forming a group of chemicals called trihalomethanes, or THM. THM concentrations vary from one water system to the next, depending on the local supply of organic junk, the water temperature, and technology. Scientists are still investigating how much THM is safe to drink for a young woman who hopes to give birth to healthy kids. I wonder, though, how THM might affect Cheeky, the squirrels, and the birds. Their bodies are so different from mine. Cheeky, for all I know, is a girl munk who'll have munklets next spring. So I'm relieved when I read that the Portland Water District has switched from chlorine to ozone for disinfection. The THM level in my tap water has zeroed out.

That doesn't mean the Cast Iron is composed of H_2O alone. In addition to the minerals dissolved off the landscape by flowing rain and snow, this piped river contains chemicals that engineers pour into it on purpose. They add sodium hydroxide to lower the acidity. They toss in zinc orthophosphate, which coats old lead pipes in people's houses and thereby preserves IQ points. And they run a stream of fluoride into the river to preserve the teeth of children—and chipmunks? Presumably. The fluoride is the only additive that raises serious concern in humans. Questions about it eroding our bones and discoloring our teeth won't go away. But as for Cheeky, he's going to confront hazards far more fatal than blotchy teeth.

More alarming than the chemicals that engineers add to the Cast Iron are the chemicals I pee into the river. Recently a series of tests have found that treated wastewater, which pours into streams, lakes, and oceans, holds enough drugs to dose a whale. Antibiotics, painkillers, epilepsy meds, spermicides, deodorants, acne cures, antidepressants, and lots of caffeine are among the products we spill into the false rivers that pass through our homes. How our doped pee impacts wildlife and plant life is just now being studied. Spermicides and antibiotics seem to devastate algae. Prozac apparently slows the development of fish and tadpoles. To take an optimistic view, maybe the algae-killing antibiotics counteract the algae-feeding nitrogen that washes in from farms. And perhaps Prozac helps the frogs take a positive approach to the news that atrazine is turning them into hermaphrodites.

❧

The third branch of my water system is invisible. It's the water that falls elsewhere, then migrates, for decades or even millennia, through fractured bedrock and gravel. Deep beneath the yard is a water world that flows on a timescale that makes a mockery of my kitchen clock. Dwelling in it are citizens who never see the sun. This is the world of groundwater. And the only way I can investigate it is to call a dowser, a water witch. Suppressing an eye roll, I dial the number.

I've already tried the rational approach to envisioning ground-water. I've connected the high points on a topographical map of the neighborhood and thus outlined my little watershed. My yard sits in the middle of a shell-shaped dent in the land. When rain falls uphill from me, some of it trickles down through the soil, then down deeper, into rock. Presumably, gravity keeps pulling it downhill, toward the ocean. But what path does this groundwater take under my yard? Does it follow the cobbled course of a buried riverbed? Does it zigzag through fractures in bedrock? Is there a roaring river under my lawn, or an Everglades-like sheet of dark water? Or just a hard desert of watertight stone?

I've read about groundwater, trying to learn its habits. But I was quickly distracted by the animals who live in it. I found an entire book dedicated to them, with a title so fun to say that it's permanently etched in my brain: *Stygofauna Mundi*. That's Latin for "groundwater animals of the world." *Stygofauna Mundi, Stygofauna Mundi*. Oodles of groundwater animals live under my lawn. There are nematodes—little wormy things. And copepods—

shrimpy things. And mollusks, ribbon worms, ostracods, isopods, even beetles. These folks live on organic crud that rain carries down through the soil, gravel, and rock, or they eat each other. The whole idea makes me itch with curiosity. I'd love to dig down to my groundwater level, look at these creatures, and size up their subterranean swamp. But I can't. So I await Roland Moore.

Roland is slight, and eighty-five years old. White hair peeks from under his tractor cap, and suspenders support his blue Dickies. He leans a bit to one side. This, along with his glasses, gives the impression that he's trying to sidestep attention. At his truck, he hands me a bundle of miniature flagpoles, which he collects while maintaining his neighborhood cemetery. He retrieves slim, brass L-rods from under the seat. The short end of each L disappears into a knot of red and blue knobs that make up Roland's hands. He's ready. He takes a position near Cheeky's primary hole. I struggle not to steer him away.

I'm a bit embarrassed to report that Cheeky has become the sun around which my world revolves. A few weeks after he scampered into my life, I was working in my upstairs office when I sensed I was not alone. Looking around, I caught the familiar blur of his fur. He had conquered the stairs. Within a few minutes he had popped from my footrest to my knee, launched from there to the keyboard tray, where he typed an apostrophe, then summited the desk. After that, I kept a seed cup on my desk. Any morning I spent there, I spent in the company of Cheeky. I'd hear his *pitta-pitta-pitta* on the stairs, and a few seconds later I'd feel his claws on my leg. He'd dive into his work, then speed down the stairs. Through the window I'd see him bounce off the deck and into the shrubbery where he hides one of his tunnel entrances. On days that I woke early, opened the door, and returned to bed, Cheeky would *pitta* around until he found me. He'd make an exploration of my head until I got up to fill his cup. But as thoroughly as Cheeky knew my home, all I knew of his burrow was what I'd read: It should be many yards long, with multiple doors and storerooms and a bedroom. It was only by accident that I discovered one of his entrances. I happened to be looking one day when his head popped out of an old posthole. This three-inch hole, lined with rocks, came with the house and had never before shown signs of life. Now that I know where to watch, I often see Cheeky emerge from it. And I avoid the area. I can only imagine what it sounds like to Cheeky when a massive primate thuds across his ceiling. But I don't tell Roland, and I don't take his elbow and lead him away. I don't want him to think I'm a nutball.

Roland's rods sweep back and forth. "This area," he says, "is an area where we don't have any primary water." That was quick. But he's not done. Primary water is just the big stuff. My yard may have veins of water that flow off an underground spring, or dome, of water, he says. Roland shifts gears somehow and starts to walk toward the back of the yard. His stride is a shuffle, his black brogans scudding across the lawn. Just a few feet past Cheeky's hole, his fists tilt and the rods swing toward each other: clink. He turns and follows the vein toward my right fence, pointing the rods at the ground when he nears the apple tree. He directs me to poke a mini-flagpole into the ground. He follows the vein to the left fence, and I stick in another flagpole. He resumes his shuffle toward the back of the yard. He doesn't get far. We mark off three parallel veins that fall about ten feet apart. Then he traces three more that angle a bit. When Roland finds himself confronting the black-raspberry canes, he announces, "That's about it." But that doesn't mean he's done. He swings around.

Now Roland is going to tell me how deep each vein is. He goes to the last one we marked and stands on it. Holding the rods out, he mumbles to himself. "What are you saying?" I ask. He mumbles louder, "From the bottom of my foot to the top of this water vein is it over ninety feet?" Clink. "Over a hundred?" The rods dip in denial. "From the bottom of my foot to the top of this water vein is it ninety-nine . . . ninety-eight . . . ninety-seven . . . ninety-six . . ." Clink.

There you have it. Ninety-six feet under my feet runs a vein of water. The other seven veins inhabit the same depth, give or take a foot. Back by Cheeky's hole, Roland aims the rods at my North Neighbor's yard. He's not sure, but he thinks the source of the water beneath my yard is a dome of water lying deep beneath the neighbor's apple tree. Right. We sit down on the deck to clarify a few things. When Roland sits, his fingers explore his knees, and his feet rock and slide as though they're feeling for something. I ask why he uses L-rods instead of a forked stick.

L-rods are best for tracing veins, he says. A Y-rod is harder. "But you can use anything," he says. "Whenever I'm driving over this way, I program my fingers, then I ask if the drawbridge is open. I say, 'Show me yes,' and my finger might bend like this. I say, 'Show me no,' and it'll bend like that. Then I ask, 'Is the bridge open?'"

Oh, dear. My skepticism was suspended by a thread while we did the water dowsing. Now the thread breaks. And Roland isn't done yet.

"Probably more than finding water now, I clear houses of negative energy," he asserts. "I get three or four calls a week for that." My face muscles fight amongst themselves. At a loss for what to say, I indulge morbid curiosity. His card, printed on glimmering, red plastic, says he does "biolocation on site and remote." I ask what *remote* means. Roland replies that he doesn't have to leave home to dowse. He can sit at his desk, draw a picture of the property in question, and program a handheld pendulum to give him answers. If he needs to divert a vein of water into a dry well or something, he might stab a little bar onto the paper, then tap it one way or another. Argh! This is going from bad to worse. I thank him, pay him extra for his time, and bid him adieu.

I mock him that night among a group of friends. My hostess gives me a look, then goes to fetch a handful of pendant necklaces. "Just let it hang until it's still," she says, giving each of us a necklace. "Now ask it, 'Show me yes.'" Snickering, we do. The silence expands as each pendant begins to swing. "Now ask it to show you no," our hostess prompts. We obey. The pendants quiet themselves, then swing in a new direction. I can't explain it, and I won't try.

The next morning, curious about how Roland's water lines compare to the underlying geology, I unroll my bedrock map. The bands of color show how the layered bedrock under my yard was folded, accordion style. The colored bands run in the same direction as the water veins Roland marked.

If Roland is right, my dream of digging a pond for the backyard creatures isn't going to come true. I'd like Cheeky and the gang to have natural, earth-cleaned water. But calling in a well driller seems excessive, even for someone who converses with crows and kisses chipmunks. And even if I did install a cistern to collect rainwater from my roof, this year's harvest wouldn't wet a woodchuck's whistle. By September, groundwater levels across the state will be at a record low. Streams will be running dry. My backyard companions, like me, will have to settle for the Cast Iron River and all its chemistry.

SUMMER

4

I LOVE YOU, NOW SPIT OUT MY AZALEA

SOMETHING'S UP. THREE squirrels are sharing one apple tree. And I hear no cussing, see no bloody tails flutter to the ground. Something is wrong here. As I creep close, three gray bodies scramble down various branches and disappear into holes in the trunk. After a moment, two eyes appear in one hole. In another hole, two noses emerge, cheeks pressed together, eyes rolled my way. The cheeks are a bit chubby. The noses are dark brown. They've got to be youngsters.

Whiskers twitch, and the pair ooze in tandem from their hole. They plop onto a branch and gaze at me. The third remains in his low hole, looking sour. His siblings turn to each other for a bit of grooming, or playing, it's hard to tell which. One bear-hugs the other's head. The huggee slips free and grabs the hugger's tail, running it through his paws and jaws. The tailee twists to hug her sibling around the middle, and the huggee slithers around the hugger until he's the hugger now. The cuteness is immense. And a bit clumsy—in fact it won't be long before little squirrel bodies litter the streets all over town. Maybe the churlish squirrel—I'll call him Earl—has a premonition of the hard days ahead. He props both elbows on his windowstill and settles in for a protracted glare. I'm sensitive to his discomfort, but I can't pull myself away. His siblings are practicing their footwork in the apple-tree canopy. As one makes a dash up a flight of parallel branches, her foot misses a grip and she slips. She descends and starts over again. My own pegs are a bit stiff from standing still, and I shift my weight. Everyone dives down a hole again. Soon they're back, clambering through the leaves. Well, not Earl, who resumes his glaring. But the other two eat leaves, chew bits of bark, and wrestle as though I've vanished. They slip on branches and catch themselves. They jump onto each other's heads. They lie flat as though fatigued and bite hunks of bark off the tree. At the risk of repeating myself, the cuteness is absurd. Even Earl starts to relax, his eyes closing and his head nodding. He rouses himself and retreats. "Chirrrrrrr," he snarls as he goes. His siblings cock their ears for a minute. Then they're wrestling again.

In the coming weeks the kids will fan out from their tree. They'll hop too slowly across expanses of grass, exposing their spines to predators. They'll dawdle in roads, acquainting themselves with motor vehicles. And flicking their tails in my direction, they'll climb the deck to ogle the tender corners of my house. Already, someone is spreeing in the wall at night. It sounds too small to be a squirrel, but that's scant comfort.

My personal goal for my personal yard is emerging: I want all native creatures to be welcome here. But my goal hits little obstacles. Little, furry obstacles that eat my home and pee in my ceiling.

Disregard for humanity is a helpful quality in a mammal who aims to thrive in town. Squirrels have buckets of that disregard. They have to focus all their regard on getting through a year alive, and with their tail intact.

Squirrel life looks carefree, at first blush. But I think we may mistake one squirrel for the next and falsely conclude that they're always around, and always finding food. From the moment I began watching my squirrels, I saw tragedy and high turnover. Right away I noticed that one animal was missing all but a tuft of tail. Stumpy, I called him. Without the flag of tail following him across the lawn, his gait resembled that of a mini-rabbit. I wondered what had happened to him. A birth defect? Not likely. In the definitive squirrel reference, *North American Tree Squirrels,* I read that ten percent of squirrels have suffered a serious tail-shortening experience, almost always provided by one of their own kind. The males, when chasing a female in heat, tend to run in packs. And the guys in the back distract the guys in front by snipping off their tails.

It didn't take me long to realize I had bestowed the name Stumpy too hastily. I was sitting in the oak tree one day when I heard a rustle in the maple sapling next door. A squirrel was descending into the Bamboo Wilderness. She was not doing it gracefully. She slipped, scratched, hung by one foot. In jerks and hitches she clambered onto a fallen log. And there I could see that she was missing most of her tail. She had an inch more than Stumpy, so in my notes I referred to her as SemiStumpy. She came toward me. At ten feet she spotted me. She froze, and I did, too, hoping she'd lose interest. As I watched her watch me, it came to my attention that she wasn't really a SemiStumpy at all. She was Double Stumpy. She was Seriously Stumpy, and Stumpy Squared: She had no right rear foot. That explains the messy descent. I take back Stumpy's name, and give it to this new girl. After freezing for five minutes I hit the limit of my endurance. Things itched. Other things ached. Stumpy had not moved a molecule and showed no discomfort. This freeze response works. Many predators don't pay much attention to stationary objects. I once watched a dog rush around the yard, frenzied by the scent of woodchuck, while that woodchuck sat like a stone, smack in the middle of the lawn. The 'chuck could have trimmed the dog's toenails, it passed so close.

I need to either move or scream, so I talk softly to Stumpy as I shift. She thaws. She turns to face me, revealing a left ear that's missing a V of flesh. She raises her stumpy leg to scratch an itch, but makes no contact. She's calm enough to get back to her chores. Her furry peg leg slips on the wood as she hops along the tree and enters the Bamboo Wilderness. I hear her climb a sumac, then hear a scratch and a plop as she falls. She proves to be a regular in the yard. Her hop is high and short, and she's often chased by other squirrels. She's nervous and quick to rush for a tree when she hears the alarm call of a blue jay or a crow. But come fall, she'll get the best of her brethren, bouncing away to a new territory with their acorns in her teeth.

Few scientists study urban mammals—the woodchucks and possums and chipmunks. These creatures must lack economic and ecological importance. But squirrels are studied. It's not because so many people love or hate squirrels. Rather, squirrels do interesting things. They stash food, then locate it again. The acorns they misplace keep oak trees sprouting. In some ecosystems, squirrels are so enmeshed with the reproduction of pines and oaks that without them it's possible the ecosystem could collapse. Last, but definitely not least, squirrels are easy to study. They sleep at night, which biologists also prefer to do. And they're so casual around people that much of their behavior can be observed. The result is that biologists are able to study a mammal with more ease and comfort than normal.

🐿

Under the pressure of scientific inquiry, the squirrel's mind is revealing its secrets. I find the revelations detailed in *North American Tree Squirrels,* which relates the peculiar behavior of not just squirrels, but also the people who study squirrels. Scholars Mike Steele and John Koprowski describe squirrel experiments with a sharp eye and a tongue in cheek.

Consider the mystery of squirrels' resource management. When presented with acorns from red-oak trees and white-oak trees, squirrels do different things with each. Why, squirrologists inquired. And how do they tell the acorns apart?

Enter the other nut-smiths, Steele, Koprowksi, and other researchers who tamper with acorns to test squirrels. The first task is to confirm that squirrels do treat red and white acorns differently. Steele and Koprowski fix tiny metal tags to both kinds of acorns and give them to squirrels. The squirrels eat some. And they bury some. The nut-smithies fire up a metal detector and retrieve the buried acorns. Conclusion: Squirrels eat the white-oak acorns and bury the reds for winter.

Why? The researchers are confident they know: White-oak acorns germinate the same autumn that they drop from the tree.

Burying them would be a waste of squirrel time, because they'd promptly sprout. Red-oak acorns, however, don't germinate until the following year. They're ideal for storing. Squirrels have evolved an instinct to store acorns that have a long shelf life.

So, how do they know the difference? Researchers doubt that squirrels simply eyeball their acorns. First of all, the rodent's wide-set eyes aren't suited for close work. Second, when a squirrel grasps an acorn in his paws, he rolls it under his nose, rather than peering at it. So the scholars tackled the smell issue. It took three research seasons to build the prefect experimental acorn, or fakorn. The shell had to be sliced clean in half; the cotyledon taken out and ground to powder; then the shells restuffed with a custom nutmeat mixed from ground cotyledon and various additives. The investigators put red-acorn meat in white shells, white-acorn meat in red shells, and even adjusted the ratio of bitter tannin to yummy fat in their fakorns. To remove any chemical clues from the shells, they soaked them in acetone. Then they glued the shells shut with odorless adhesive. The squirrels were stumped. They couldn't tell the difference between fakorns stuffed with red meat and those stuffed with white. Only when researchers omitted the shell-soaking stage did squirrels respond. In that case, they buried any kind of fakorn that was presented in a red-oak shell. They ate any fakorn in a white-oak shell. So the smell of a red-oak shell cues a squirrel to bury, right?

Wrong. The final round of this investigation asked whether the squirrels could detect if an acorn was preparing to sprout. And they can. When researchers handed squirrels two red-oak acorns, one dormant, the other winter-primed for germination, squirrels nipped the embryo out of the reds that were preparing to sprout, then buried them. The scholars had seen this before: If a squirrel is given only white-oak acorns, he will bite out the embryo that causes the acorn to sprout and bury the sterilized nut. Scientists had just never seen it done with red-oak acorns. So, eureka: By sniffing the shell of *any* acorn, squirrels can smell fertility. And they nip it in the bud.

I must say, biting out the embryo from a seed is one of those behaviors that give me pause. When a squirrel bites out the tip of

73

an acorn, he is not thinking back to what he learned about germination in his botany class. He's relying on a knee-jerk instinct, which some distant ancestor got by way of a genetic mutation. The bite-out-the-embryo instinct caused such success in that ancestor's family that his descendants thrived to the exclusion of all their relatives. The delicacy, the freakish precision, of evolution sometimes floors me.

Another excellent question is how squirrels recover their stashed acorns. As soon as the red oak in my backyard began dropping acorns in late summer, the squirrels were spiriting them away. Actually, they begin harvesting acorns the moment they swelled into existence. In late June, squirrels young and old were thrashing around the canopy, pruning and dropping entire twigs in their quest for baby acorns. As the acorns grew fat, squirrels harvested them directly, carrying them down the trunk and into the bushes. By the time nuts were pelting like hail on the shed roof, the squirrels were in a burying delirium. They might carry an acorn just a few feet before digging the quickest hole, tamping the earth, and pawing the grass straight. (The crows do the same thing when there's a lot of food lying around. Their first priority is to get the goods out of sight. After that, they can move it around at leisure.) As the hail of acorns diminished, the squirrels shifted their attention to cache management. This was an endless task. I watched my squirrels psyche each other out by pretending to bury the same acorn in three different places. Where the nut ends up, only the burier knows. In addition to handling their own nuts, I see my squirrels trail the crows and blue jays, nosing under leaves and pawing aside grass to see if the birds stored anything worth stealing. Many times I watched Stumpy stealth-walking into my yard from the North Neighbor's yard. She would cross my lawn and creep into the Bamboo Wilderness. A few minutes later, she would emerge, acorn in mouth, and run for home. Sometimes the injured party discovered her and gave chase with much clattering of teeth. But time after time, Stumpy sprinted home with the prize.

How do squirrels find buried food? From what researchers can deduce, memory and odor both play a role. Squirrels can find nuts

74

they've buried after fakorn-makers deodorize them, which implicates memory. And they can also find nuts their neighbors buried, which implicates smell. As to their overall efficiency at finding what they've hidden, studies are few. But the *Tree Squirrels* authors say it can reach ninety-five percent.

The more I learn about my squirrels, the more I'm mortified by the dangers they face. And the dangers become clear. Soon after my young cuties pop out of their apple tree, young cuties appear all over town, flattened. Some have been hit by cars. Some may have been chased into traffic by dogs or cats. Others probably fell from trees, which happens more often than I would have thought—Steele and Koprowski report that nearly one in twenty gray squirrels harbors a healed fracture in one leg or another. The flattened squirrels reenter the food chain, at least. In my neighborhood, crows, perhaps chuckling at the memory of squirrels swatting them, tug off strips of red flesh and store them in the trees.

In spite of what looks like an epidemic of street fatalities, starvation seems to be the most common cause of squirrel mortality. The details give me a new admiration for the mother that brought up my three cuties. Last summer and fall she stashed and stashed, ate and ate, putting on fat. As the air grew colder, she grew less choosy. Instead of eating the mild tops off acorns, she chewed each nut to its bitter end. The early darkness sent her up to her drey, a hollow sphere of leaves she shared with her female kin. Winter struck, and still she went out to feed, digging through snow and frozen soil to recover a meal. In January she mated and forty-four days later lost one tenth of her body weight in the form of pink progeny. Her fatty milk depended on more acorns, and out she went to extract them. As the snow piled up in my yard, I noticed that my squirrels spent a lot of their excavating time in wind cups that formed in the lee of trees. I wonder if they buried more acorns in these spots, or if these were the only places they could hope to dig to the ground.

The three kids developed a taste for solid food just as spring brought out the new leaves and flower buds. They had survived. If a late frost had killed this first round of leaves, that might have been the end. Instead, it looks like the end for my neighbor's old apple tree that hosts the adorable brats. The squirrel babies have eaten the bejesus out of it, and they're ready for something new. With their brown velvet noses testing the air, they stealth-walk toward my home.

Hang on. Now that I think back, Mom Squirrel didn't face much hardship at all when she was nursing her brood. She had my birdseed to fall back on. For the umpteenth time, I had hung a bird feeder full of sunflower seed. For the umpteenth time, squirrels had giggled in their sleeves, shinnied up the metal pole, and emptied the feeder. I apologized to my wallet and slapped down twenty bucks for a tubular metal baffle that slid over the pole. Oh, it was gratifying to watch a squirrel shinny up the pole and halt with her head under the baffle. Hee, hee! The three days it took for them to crack this particular nut were delightful for me, but in retrospect not worth twenty dollars. On day four, one determined squirrel leaped straight up past the tube and caught the perch-rail of the feeder. I moved the feeder up four inches to the highest part of the pole. This bought me one hour. I expect if I could raise the feeder an inch a day, I'd eventually have a squirrel who could medal in an Olympic jumping event. So I gave up. The truth is I felt sorry for the animal anyway, knowing it could be a nursing mother whose cupboard held only sour old acorns. I took to throwing sunflower seeds on the warming ground. This produced some rousing fights. Whichever squirrel was eating first seemed to have the right to throw the others out. And he who was evicted was often so incensed that woe betide the spring blossom who might witness his humiliation. One frustrated fellow barreled over to a tulip and bit its stalk in half. Another day, a loser ran to a daffodil, pulled its head to the ground, and slashed its belly with his hind feet. They were gory times. But as summer unfolded, maple seeds fell, acorns swelled, and the tension dissipated.

Considering my generosity, I grow tired of the squirrels' distrust of me. I feed them and talk to them, but they treat me like a serial killer. When I step outside, they race for the safety of the trees, their tails flailing. The crows, whom I've been talking to and feeding for months now, are just as bad. They'll tolerate me sitting on the deck, but they still scatter if I *look* at them. Jeeeez. Obviously, these critters are lousy judges of character, but it still makes me feel bad. The companionship of Cheeky the Chipmunk is all the more gratifying, in light of the bloodcurdling effect I have on the crows and squirrels. While the squirrels flick their tails on a high branch, Cheeky sits in my palm. I rub behind his ears and run his copper tail through my fingers. I poke his cheek and hear the seeds shift inside. I rub my thumb over his face and he shuts an eye. Only when I close my thumb and forefinger on either side of his distended cheeks does he react. He rears back and looks up at my face as though he's never met anyone so rude. Then he resumes shoveling in the sunflower seeds.

Why do I yearn for the affection of the animals in my yard? Maybe the price of being the top-dog species is that we scare everyone else and then feel isolated. Maybe this cross-species ardor is a side effect of being a community-minded animal with a complicated brain. Who knows. Beyond dispute is that people crave animal companionship. And by a country mile, the favored companions are mammals.

My biophilia even extends to woodchucks, now that I'm not trying to grow vegetables. At my previous house the wretches trundled from beneath the shed at dawn and mowed my peas to the ground, day after day. I hardened my heart against them and loosed my hound, a dog who has since passed on. From time to time, I would be rewarded with a fresh corpse buried in the flower bed. One day I watched the hound stand on his side of the fence and snap the necks of four baby 'chucks who ambled, one after the next, out of the neighbor's tall grass and between the pickets. What a glorious scene! When the neighbor girl caught on and wailed, I sternly ordered my dog indoors, where I fed him hamburger. But at this new house, I look forward to seeing Big Fat Momma chug out of the bushes. She must weigh ten or twelve pounds, and when she props herself on her butt to scout for danger, her fat and fur drape her like royalty. She selects broad leaves from my weedy lawn, and I watch them waggle into her jaws. She, like Cheeky, tolerates me talking to her. She maintains a network of escape holes, and if I should become impertinent, she knows where the nearest one is.

Now that I'm studying the ecology of my yard, it's dawning on me that woodchucks are real, native animals who don't deserve to be hounded. The woodchuck's wild-animal status was driven home again when I came across a reproduction of a bill of goods from an old trading post. The prices of goods (shirt, shirt with ruffles, tobacco, etc.) were listed in animal skins. And to my surprise, three hundred years ago you could buy more shirts with a wood-chuck pelt than with a beaver pelt. That must mean that they were more rare. That puzzled me, until I remembered that the land was almost entirely forested until Europeans got serious about logging and farming in the 1800s. Although woodchucks will use forest for winter denning, they do their summer eating in open fields. The European pastures, and the backyards that evolved from them over the past century, were a woodchuck's dream come true. Their population must have exploded, along with the population of cows and horses. Now they're anything but rare, and most people asked to place a value on them would name a number deep in the red.

Squirrels and Cheeky aside, most mammals who do well in suburbs and cities do their thing at night, when people and dogs are less likely to bug them. I'm not keen on spending my nights among the mosquitoes, waiting for animals to come out and exhibit their natural history. Fortunately, there are other ways of getting the lay of the nighttime land. One is the roadkill survey. A late-winter road with black-and-white carcasses reminds me that skunks are rousing themselves in preparation for mating. If it's late spring and the roadkilled have ringed tails, I'm reminded that year-old male raccoons are dispersing from their mother's territory to make their fortune. It's nice to know that roadkill isn't a total waste of life. It can be educational.

A less gruesome means of counting noses is to set traps. I head into the yard with a bucket of playground sand and spread it every place where animal traffic has worn a ditch under the fence or pushed up the wire. I also set up a scent post. (This is a shallow aluminum tray filled with sand. Jabbed through it and into the ground is a dowel. Atop the dowel is a blob of the universal attractant, peanut butter.) With the sand traps, every morning is like Christmas. I scuttle out to see what I've caught. Unless it rained in the night, there's always something in the traps.

Crouching to put the sand between me and the slanting sunlight, I see topography in the sand. A long hand with claw holes at the end is a skunk's front foot. Also clawed, but with a ridge between the pudgy heel and the pudgy ball, is the skunk's hind foot. Two rows of tiny scuffs with a snaky tail track in between record the crossing of a mouse. A tight ring of small craters is a cat track. A slug's slime trail has curled into a sandy tube. Bird feet make delicate pitchfork shapes. A thread-thin line of minuscule dents marks the progress of an insect. The longer I look, the more I see. The rusting fence has snagged a hair. It's long and multicolored, like one of the stray cats who stalks my birds. Now I notice an aging feather, then another, then white bones matted with more plumage. An old bird kill. At the scent post the tracks are all over each other. Squirrels, whose hands are larger than I'd have guessed, were up early. One of them I can identify with certainty: Stumpy's peg leg left round dents in the sand.

Not every nocturnal is shy. If I left my door open at night, I think the skunks might come right in. One evening a dog that I'm babysitting growls out through the dining room screen. Beside the deck steps, a bucket is rocking. A trowel clangs. I step outside, leaving the light off. Looking over the railing, I see at first only white blots in the flower bed. As my eyes adjust, I can make out two tiny skunk kits wrestling like, well, kittens. They untangle themselves and mosey through the plants, their airy tails standing tall. At the corner of the deck, one turns left to visit the neighbor's dog. The other putters along in the flowers, a tad jumpy without his sibling. A screen door slams down the street, and he hops stiffly around to aim his tail at the noise. He thinks better of it and shuffles back to point his ears at the sound. His face is a triangle with glittering, black eyes. A white stripe runs up each flank. His legs are stiff, giving him the locomotion of a stuffed toy. A leaf crackles in the Bamboo Wilderness, and he hops around to face that . . . no, no, better to aim the tail. I'm kneeling on the deck now, and he's directly below me. "Hello, little Stunky." He peers up. Blinks. Then he noses through the leaves and noodles away down the fence line.

Later in the summer I meet one of these fellows again. I'm sitting in the dark listening for mice and feeding the mosquitoes, when a black-and-white blot the size of a football bobbles past in the grass. Silently, I get up and follow. He moves slowly, nose to the ground. Skunks have weak eyesight, but good noses and strong claws that rip Japanese beetle grubs out of the ground. He stops and digs hard. While his head is down, I move in for a closer look. He digs again, and I move to within ten feet. He digs by the shed and turns himself around. Now, absorbed in his work, he's meandering back in my direction. He's eight feet away. Then six. I'm half-crouched to see him better, I'm being perforated by mosquitoes, and I am in a quandary. Should I let him know I'm here? Or might he go past without noticing? An adult skunk can fling yellow oil ten feet. I've seen one fling it over his own back and into the face of my dear departed hound. The guy coming my way is only half-grown, but I'm sure he's capable of slagging his sulfur six feet. A friend who got skunked once told me that you burn your clothes and prepare for two weeks of strangers in checkout lines griping about an odor.

"Ahem," I say softly. The skunk raises his head and stares. "It's okay, Stunky. Put your tail down and no one will get hurt." He isn't convinced. He stamps his back feet. "Oh, no need for that. Nice skunk. Pretty skunk." What am I wearing, and how will it burn? Stunky's head weaves from side to side, and then he makes a break for it. He turns and stiff-legs into the tall grass by the shed. He stops with his tail sticking out. "Thank you, Stunky." The tail disappears. I swat seventeen mosquitoes. In a moment Stunky

80

emerges from the far side of the shed, fully recovered. He zigzags across the dark lawn, digging in the moonlight. The mulch under the raspberries crackles. Then his white stripes disappear under the back fence. They glow once more under the neighbor's lilac, and Stunky is gone.

I'm not particular about holes in my turf, and I have no affection for Japanese beetles, so I have nearly no objection to skunks. I often babysit other people's dogs, but if I turn the deck light on before letting dogs out at night, confrontations are avoided. I do, however, have an issue with skunk poo. One of my dog friends found some and rolled in it, and it really doesn't get much worse. It's just as powerful as the yellow oil, but in a far more offensive form. Still, it's an opportunity to learn about Stunky's diet, so one day I pinch my nose and collect a sample in a plastic bag. I slide it under the microscope. It's a mass of intertwined insect parts. Brown antennae curve around shards of mahogany shell. Dark mandibles pinch earwig tails, which in turn pinch wings. They're from modest-size insects, these parts. Stunky must have to eat all night.

My biophilia complex does not fire up to the extent that I want to kiss the skunk. But I do notice that if I put out food that the crows don't care for, or that is simply too terrifying for them to approach, it disappears overnight. I like to think I'm feeding the Japanese beetle eradication corps.

As much time as I spend outside at night, I hear precious few mice. They're alleged to be supernumerary in lots of ecosystems, ubiquitous as ants. But even their tracks are rare in my sand traps. It's a puzzle. One summer evening when Chuck Lubelczyk comes by to set the mosquito trap, he also brings a sack of special, biologist mousetraps. They're square tubes of aluminum with a flap door in one end. We coat gobs of peanut butter in rolled oats and drop a gob into each trap. Chuck says mice travel the edges of things whenever possible, so we set the traps against the fence, the house, and shed. "Put them all in the shade first thing in the morning," Chuck says. "Those traps get hot. And I just want to prepare you. If we catch shrews, they'll probably be dead by morning. Their metabolism is so high they literally starve to death in a few hours."

We get no shrews. When I nip out to check the traps in the morning, five have shut. I rotate one slowly so I can peek in the door without opening it wide. Four white feet appear. Another trap shifts in my hand when I pick it up. The other three are false alarms, empty. We've probably bagged white-footed mice, but deer mice look nearly identical: gray on top, white below. Both are native, unlike the brown house mouse.

When Chuck arrives, he pulls on blue surgical gloves. Mice can carry hantavirus, a rip-snorting respiratory bug. He upends a trap over a skinny plastic bag. The mouse hits bottom and heads back up, but Chuck is already clipping the bag to a scale. About two thirds of an ounce.

He reaches in to catch the mouse's tail. It's a male. Chuck lowers him onto the step, then runs a thumb and finger down his spine until he's got the mouse by the scruff. He lifts him and begins ruffling his fur with tweezers, looking for parasites. The mouse hangs quietly, silken whiskers waving a little. The ears and eyes are large, for nocturnal use, Chuck notes. He's poking behind the mouse's ear, a favorite spot for ticks. He finds none there, nor among the whiskers, a sensitive area that mice are loath to scratch. A deer tick would be especially interesting to find, in light of its role in transmitting Lyme disease. I've never seen a deer tick here before, but Chuck has vials of alcohol at the ready.

Now, here's an unexpected thing about mice. They are less likely to aid and abet in giving me Lyme disease if my surroundings support a lot of other small mammals. The bacterium that causes Lyme disease lives in lots of little critters. But of all of them, white-footed mice are best at passing the bacterium to a biting tick. So the more mice you have, the more likely it is that your local ticks will carry Lyme disease. However, if for every mouse in your neighborhood there are ten other small mammals for ticks to

82

choose from, then the odds of a tick biting an infected mouse plummet. So the rate of infected ticks goes down. And— drumroll, please—the rate of infected people goes down. At least that's the conclusion of a couple of studies, both preliminary. It sounds reasonable to me. It's one more reason to roll out the welcome mat for Stunky, the squirrels, and the rest of the furbearers.

Later, when I'm wondering what the mice have done for me lately, I stumble across another curious fact. Mice help distribute special fungal spores. These fungi grow on the roots of trees, helping them gather nutrients from soil. Mice eat the fungal spores here and poop them there, helping them spread from tree to tree. Mice serve their ecosystem in other ways, too. They disperse plant seeds when they stash them for winter. And they're a staple in the diet of small carnivores ranging from foxes and coyotes to birds, bobcats, and snakes. In my yard the adults are probably eaten by cats, while the babies may be taken by skunks and possums.

Chuck dumps out our second mouse. This one is female. "Normally, they're loaded with fleas, but these seem kinda clean," he complains. On cue, a tiny flea hops onto his glove. Like the first, this mouse hangs quietly as Chuck works it over.

"Some of them will actually start grooming, washing them- selves, while you're holding them," he says with a grin. "You just get used to that, then one lays into you." We carry the mouse to the back fence where she was caught. "You ready to go back?" Chuck asks, and strokes her head twice. Her feet hit the ground and she springs into the brush pile. "It's a completely different world at night," Chuck muses, looking around. "Most people have no concept of what's out here when it's dark."

As sappy as we humans are about a furry face, our day-to-day rela- tionship with urban and suburban mammals is often a struggle. We crave the contact, then, often too late, we see the warts. Or the ticks. Or the holes in the house. When I visited England recently, I learned that foxes roam English towns. How cute that must be! thought I. I wish foxes roamed my town! And then I got the details. The urban foxes abandon their social structure, their dawn-and-dusk schedule, and their natural diet. They raid garbage cans in daylight and live short, scruffy lives. But, boy, I bet they were cute when they first arrived, and I can imagine the rapture of the first English urbanites who set out a dish of cat food to lure them closer . . .

Anthony DeNicola, a wildlife ecologist who deals with animals who have worn out their welcome, says it often takes a decade for our feelings about a newly suburbanized mammal to progress from awe to awful. "By ten generations, the fear of man is almost gone in the animal," he explains. "And then people tell me, 'We'd love anything you could do to kill them.' It's sad. You set the animal up. You provide an easy food source. The animal habituates. And then you persecute it."

Tony is not your average Ph.D. ecologist. For one thing, most ecologists I know look as if they got their last haircut from another ecologist, who had five minutes and a Leatherman. Tony wears a crew cut and looks like a marine. For another thing, instead of laboring in academia, he runs a nonprofit company called White Buffalo. And an awful lot of White Buffalo's business is sharp-shooting extra deer in the suburbs. He also researches deer contra-ceptives.

Deer culling is a nasty business. And that's mainly because we *Homo sapiens* are so confused about how close we want to be to Nature, and about what Nature should look like. As a general rule, we focus our adoration and our money on a few creatures that scientists call charismatic megafauna. That's Latinate for "big, cute animals." Deer qualify. Accordingly, animal rights activists some-times disrupt Tony's work. If he were to spend his evenings killing locusts, I doubt anyone would protest. Yet locusts are what white-tailed deer become when they enter the suburbs.

You can't blame them for coming. People create ideal deer habitat in their backyards. Hunters don't intrude on this grassy-shrubby Eden. So the deer descend, and if people don't throw rocks at them, they eat the place bare. Batting their long eyelashes and swiveling their cute jaws, they grind and swallow everything

84

from maple trees to rhododendrons and pansies. Then they frolic into the streets to fold up like origami against the bumper of an automobile.

Personally, I'd rather have Tony drop a deer with one shot to the head than see the animal eat itself into starvation or die of ruptured organs by the roadside—no question. I guess in my ideal world, coyotes, wolves, cougars, and predatory humans would be plentiful enough to eat the extra deer, and we wouldn't have to make this choice. But the reality is that suburban deer experience an unnaturally low rate of predation. Another fact is that we humans don't like deer stripping our azaleas, or playing speed bump on our roads. Given that, shooting them is currently the only affordable solution. The deer culled by White Buffalo are given to food banks, which must make them one of the most environmentally friendly foods in America today. These free-range animals live off the fat of the land, demanding no truckloads of corn, and generating no stockyards of manure. They're local, so the meat needn't be trucked a thousand miles. If the suburbs are producing high-quality food for low-income people, I guess I'm all for it. Perhaps I digress.

There's a chance my ideal world will come to pass, giving me the opportunity to see if life with large predators is really so blissful. Those carnivores are finding their way into the suburbs in many parts of North America. I was thrilled when I came across a *New York Times* article about cougars stalking the mule deer of suburban Boulder. Even though the cougars kill pets and stalk people, the *Times* reported, many people empathize with the cats and forgive them. They don't report the naughty lions to authorities, for fear of the lion being punished. Lured by fat deer, the article noted, the cats are moving east. It's predicted they'll soon reach New Jersey.

The *Times* may be behind the times. Since cougars were hunted out of the eastern and midwestern states by European settlers, they've been a western phenomenon. As coyotes did, cougars are now spreading. For many years wildlife officials dismissed cougar sightings in my neck of the woods as either hallucinations or sightings of released pets. And in many cases, they were probably right. But when I locate a map showing confirmed cougar sightings in the East, I simply can't believe they all were pets. Some of the animals were seen with kittens, which seems highly un-petlike. Wildlife officials are reconsidering the situation, although a prudent investigator would have to admit we may never know for sure where these cougars are coming from. They are coming, however.

When I click on my own state's map at www.eastern-cougarnet.org, I find five dots. An inland spot marks the place where a hunter watched a mother and kitten for twenty minutes, then officials later confirmed the tracks. A midcoast spot denotes two more confirmed sightings. Way up at the Québec border a hunter said he watched a mother and two kittens; wardens later classified the tracks as "probable" cougar prints. And what's this circle right on top of my house? I had forgotten: In 1995 a woman in the town next door saw a cougar drinking from a pond. DNA testing on hair from the site confirmed her diagnosis. It shouldn't surprise me. Cape Elizabeth is one of those towns that has suburbanized just enough to have large, luscious lots, and little room for hunters. The result is typical. One out of three car accidents in Cape involves a white-tailed deer. The town's few remaining farmers are battling the animals with tooth, claw, and electric fence. And the predators are coming. Cape Elizabeth has what some would call a coyote problem, and what I would call a good start. And it has definitely had at least one cougar.

Not living in a western suburb, not confronting predatory eyes in my driveway at night, I admit that my opinion isn't worth much when it comes to the joy of living cheek by jowl with large meat-eaters. And I'll confess that I'm not suited to living in fear. In the long term it gives me indigestion, and in the short term I'm prone to exactly those panicky squeaks and lurches that make a big cat's heart soar. But regardless of what I think, they're coming. The chubby deer are attracting them, and the people aren't repelling them. For the first time in many centuries, people aren't shooting every wolf, coyote, and cougar they set eyes on. For the first time in many generations of these mammals, it's fairly safe to be near people. So now we're going to find out how safe it is for people to be near these animals. Maybe, if we can control our collective squeaking and lurching, they'll help clear the streets of extra deer.

And if returning predators display an appetite for Big Fat Momma and Stumpy and Cheeky? All's fair in love and war, I suppose. If I had to choose, I'd much prefer to see Cheeky's copper pelt swallowed by a wild cat than a house cat.

5

THE ARMY OF EARTH MOVERS

EEK! WORKING AT my desk, I didn't realize Cheeky the Chipmunk was around, until I felt his claws on my bare thigh! I jumped, which made him skitter behind a pile of books. But he's a resilient guy. He springs to my lap again, then hops onto the desk, where his seed cup awaits. In he goes. To keep him acclimated to human movements, I lift the cup and slowly spin my swivel chair. He thinks that's weird, but not so weird that it spoils his appetite.

It's bad enough that I spin the chipmunk and sweet-talk the woodchuck. Today I developed feelings for a slug. If this continues, I'll be unable to step outdoors at all, for fear of crushing a native mold spore. I have to admit, even before meeting the slug, I had this sort of concern. Each time I sit in a lawn chair I cringe to think how many little citizens those plastic feet are crushing. Crossing the lawn, I'm hyperaware of my own two feet: What did I kill this time? What now? It matters, I think. My squirrels and my Cheeky depend on healthy plants. My birds need insects, which also require healthy plants. Healthy plants, in turn, need good soil. And good soil is busy soil. A slug may be the gardener's nemesis, but like the umptillion other species of creepy-crawlies in my yard, slugs improve my soil. They and the other denizens of dirt help my soil to hold a title I never expected it to win: Suburban soil, despite the insults we heap on it, may be in better shape than the soil at the neighborhood farm.

Or it may not be. If I were to draw a line from the heart of downtown, out through my suburban yard, across a farm field, and into the forest, I would probably find that some soil features improved with distance from the city. Scientists who conduct these

89

transect studies find that the concentration of poisonous metals tends to diminish as you move away from the city. The number of helpful fungi climbs. And the number of troublesome foreign invaders decreases. But for all its flaws, suburban soil does some tasks expertly. In some regards, converting farmland to lawn is an ecological improvement.

To get acquainted with my soil, I set about digging holes. I also tie red ribbons around a few items, to see how long it takes them to return to the earth. I mark a fresh oak leaf thrown to the ground by squirrels, a young acorn, and an apple from one of the trees. One of my experiments is terminated by a squirrel within a couple of hours, leaving the apple and the leaf.

My first few "holes" are really aboveground. Glamorous soil-makers like the slug do their composting on the surface of the earth, rather than within it. So I plop myself on the chaise lounge (killing oh-so-many citizens) and stare into the gloom beneath the oak tree. There is my ribboned oak leaf, brown before its time. Its surface is still shiny and smooth. It lies on a bed of last year's leaves. I pick up a stick and flip up a layer of the leaf litter. It's a thin layer indeed, just one dead leaf thick. Finer litter below is composed of gnawed baby acorns, plus shreds of grapevine and oak bark. But that layer, too, is less than an inch deep when I poke it. That's curious. In a forest, the leaf litter would be four or five inches deep. It would protect insects and soil-makers from weather and predators. Where did all my leaf litter go?

Earthworms ate it, that's what I discover. Maine is supposed to be an earthworm-free zone. In fact, all the northeastern states and Canada would be earthworm-free, in a people-free world. When successive glaciers scoured northern realms down to bedrock, they shoved both soil and soil-makers south or froze them into oblivion. While much of the country has native earthworms, every one in my yard is from immigrant stock, just as I am. His'n'her forefathers'n'mothers (earthworms are hermaphroditic) journeyed to North America in ships. Some may have arrived in the muck of a horse's hoof. Others may have been shoveled into a ship's hold with European rocks and soil that were used as ballast. Still others may have arrived with shipments of olive or apple trees, or the tea rose that one of my ancestors dug out of her Old World garden.

It's hard to imagine who could complain about an invasion of foreign earthworms. They're worshiped by gardeners, because they aerate soil and convert dead plants into organic fertilizer. But there are people with a bone to pick with worms. And the primary complaint (thus far) is that the invaders convert the dead plants to fertilizer at an alarming rate. This causes a change in soil chemistry. While that change is welcome in the lawn and garden, it's dangerous for forests. And worms are spreading into the woods. Margaret Carreiro, a biologist at the University of Louisville, compares northeastern earthworms to foreign zebra mussels that are causing billions of dollars in damage to water pipes each year as they spread through lakes and rivers in the eastern United States. "Everyone worries about zebra mussels," she has protested to *American Forests* magazine. "What about earthworms?"

Here under my oak tree, the earthworms have been diligent. Beneath the thin leaf litter, the fine litter blends into dark soil. But, as researchers have noted in real forests, the litter in my miniforest appears to be devoid of fungi. In a forest primeval, nets and mats of fungi would do much of the decomposing. They'd take years to digest the flesh of a leaf. At just the right speed, they'd convert the leaf to a form of fertilizer ideal for trees. This circle—tree feeds fungi feeds tree—was forged through evolutionary trial and error. It isn't responding well to the worms, who produce the wrong kind of nitrogen. To my eye, the oak tree looks fine. But what do I know? I thought earthworms belonged here.

As I tinker with the litter, something moves. I realize I'm staring at a jerking hair. It's two inches long. One end is white and kinky; the other is amber and straight. A squirrel hair? An ant is tugging on it. That ant gives up and paces around in the litter. A sister ant takes a turn tugging. They tag-team for twenty minutes, causing something in the fine litter to heave, but making no real progress. Impatient, I grip the hair with my tweezers. Yuck, yuck, yuck: The hair is stuck to a gluey worm, whose chewed body is clotted with more bits of litter. Yuck. I return it to the ants and move on to the next heaving pile.

I'm relieved to find this one hides a healthy sow bug. Like earthworms, these creatures convert dead plants to soil. One paper I read says that this family of creatures may produce about 140 pounds of micro-manure per acre, each year, in a French oak forest. That would translate into 28 pounds of fertilizer spread over the space of my yard, if I have a similar population. Like the earthworms in my area, almost every species of isopod (aka sow bugs, pill bugs, roly polys, wood lice) around here is imported. In addition to being foreign, they're strange. They're crustaceans—like lobsters. They breathe through gills, which limits their travels to places that are damp. They stick to urban and suburban realms and don't stray much into the wilds. Most bizarre, though, the mommies carry their eggs, and then their fresh-hatched babies, in a pouch.

I visit with an expert on soil creatures at Johns Hopkins University who tells me how it works. "The mother produces fluids to keep the eggs moist," explains Katalin Szlavecz, an angular Hungarian woman. "For a scientist, it's very convenient—you can see how many eggs each one has. And it's really cute! They have little flaps that open and"—she swishes a hand—"out they go!"

Katalin Szlavecz guides me through her lab. The shelves are crowded with bottles of preserved squirmies—worms, bugs, grubs. "Lots of dead animals," she sighs. But her workbench is loaded with terraria, and they hold living animals. A glass box of isopods has a tiny portrait of Mona Lisa taped to it. The next looks empty, but Katalin reaches in and scoops out an inch-long millipede. He's the color of varnished mahogany. "This one is just a pet animal," she says, smiling at him. "Oh, I'm sorry!" He slips from her hand and hits the floor with a click. "See," she tells me, setting him back in his tank, "I do care about them. I'll give them some food, too." From a bag of old leaves and twigs under the bench, she selects a handful of aged squirmy-chow.

"Oak leaves have to be overwintered before an isopod will touch them," she explains. "Leaves have a lot of cellulose and lignin that are hard to digest. Bacteria and fungi soften it up, especially for smaller animals that don't have strong mouth parts. They remove the tannins and break the molecules into small pieces."

On another shelf she keeps a collection of leaves in various stages of decomposition. One has been scraped only on its tender bottom surface, leaving a leaf of half thickness. Another has been cleansed of all flesh, leaving all the fine veins. "Maybe a small worm, or a springtail," Katalin guesses. The final leaf retains only its major veins. "An isopod or millipede."

Next she takes me home to her backyard and tears apart a woodpile. "We have to get rid of this pile," she says sadly. "Termites. We'll keep a little of it." She is melancholy because the woodpile is crawling with pets, easily overlooked animals making dirt. Barehanded, she pulls the bark off a log, revealing a layer of coffee grounds. "Poop. Animal poop." An orange millipede wriggles for cover and a sow bug squeezes into a crack. Katalin goes after it with a gentle fingernail. I think I hear her say, "Hey, hon," to the isopod. Among the logs, spiders guard egg sacks. Millipedes and centipedes writhe in profusion. "These guys grind up big chunks of leaf or wood," Katalin says. "By producing feces they increase the bacteria, because the feces have more surface area to work on. So then the droppings themselves are decomposed."

She sits back and looks at the logs, which are turning to soil before our eyes. "There's so much variety. There are so many ways of making a living. I don't think people are aware of it. That's why they call soil 'dirt.' I never call soil dirt. It has a negative tone to it." She smiles self-consciously. "Then, if people do know about these animals, they find them downright ugly."

Back in my own yard on my chaise, my gaze drifts through the grass to where a dandelion bud bobs. Looking closer, I discover a golden slug wrapped around the green and yellow knob. He's noshing. His tongue is too small for me to see, but I've read that it's prickly as a file, and it scrapes plant flesh into his throat. Moisture in the food is converted to moisture on his hide. I scoop him into a ceramic ramekin—it's a high-rent petri dish, but I can't find a place that sells real ones in lots of less than a hundred. I stick the slug under the microscope. He's gorgeous. He glistens in the sun. Foods come to mind—butterscotch syrup, figs. A breathing hole in his flank blows a bubble. Checking on him with the naked eye,

I'm horrified to find him looking sandy and dry. A yellow scum is forming around him in the ramekin. Egad! I've killed a hundred slugs in my life, but now that I've met this beautiful fellow, I feel terrible about his plight. I hustle him into the shade, and let him go. Perhaps addled by sunstroke, he plows under a hoop of rootlet and gets stuck. Slugs can't move backward. I feel guilty, but not guilty enough to rescue a possibly deranged slug with my bare hands. And when I investigate his natural history, I find that he's yet another urban specialist imported to this continent by people. Some slugs are native to North America, but not this common gold one. Is no earth-eater in my yard a native? Earwigs? Nope, not the common kinds. At least they prey on destructive insects, in addition to plants. Surely some of the fungi must be aboriginal.

My next "hole" is even higher off the ground. The recycling of organic matter often begins before a plant even succumbs. My crows in their doodling snap twigs from the apple trees and drop them to the lawn. The squirrels nibble off bark. Sapsuckers and beetles also bore into trees, showering shavings into the grass. Fungi, which I've been told web across the surface of every living and dead thing, erode dead bark, wood, and leaves, softening the fibers. Lichens convert the bark of my aged apple trees to crinkly lace. Then they shed fragments of themselves.

On a nearby stump I find a virtual island, on which soil-making is the main industry. A wood fungus, ruffled like a Victorian collar, is sucking nutrition from the stump. Dull at first glance, under the microscope a piece of the mushroom is marvelous. Bands of gray, cream, and brown are made of parallel fibers, and each color has a different texture. The gray rings are made of upturning fibers, for a fake-fur effect. The silver ones lie flat. The brown fibers look rotten, but they're not. I wash my hands carefully when I'm done playing with it. Later I read that this mushroom—*Coriolus versicolor*, or turkey tail—has a long history as a medicinal plant, whose antitumor reputation is being endorsed by modern science. I'm pleased to have it around.

A second fungus lurks inside the same stump, this one shaped like black fingers. When I touch it, a finger snaps off. Under the microscope, it's scary, not lovely. Black as lava and glittery, it seems constructed from a foam of black balloons. These must be spore hatcheries. Some of the bubbles are open, their spores having flown the coop. Through the matrix crawl mites the color of raw oysters, nearly too small to see. I find one more fungus, more alarming than the last. It's a foam that starts on the stump, then oozes over the grass. It's the color of toasted marshmallow on the outside, gooey white inside. Under the microscope the crust looks like toasted coconut.

All these fungi, aided by weather and a cast of thousands, have digested enough of the stump that its nutrients are available for new plants. This I know because a spray of black-raspberry canes soars from the stump's center. A bird must have deposited seeds here, and the seeds found sufficient soil to construct a new plant. Behind ramparts of dead wood, the sprouts are safe from the lawn mower. I imagine that in ten years the remainder of the stump will be colonized, leaving only a green hump in the landscape.

I wander on, into the raspberry tangle, an epicenter of plant malice. Black-raspberry canes twelve feet long anchor themselves in the soil at both ends, forming thorny leghold traps for people. Hidden beneath them are carcasses of strangled trees. Green cords of Asiatic bittersweet bind all together. I've dallied in any number of tropical jungles, and I've never lost as much blood as I've shed here. I enter clad in canvas and leather. While I'm in here, I'll prune out enough of the chaos to make berry-picking a nonfatal activity.

Starting at the edges, I reach through the thorns to snip out cane after cane. As the thicket thins, soil appears under my feet. Signs of life turn up, too—human life. When things fell in here, people dared not retrieve them. The metal ring of a Frisbee-style toy has taken on the exact color of dirt. Only its perfect circularity betrays it. When I try to toss it on the lawn it sails into the Bamboo Wilderness, where it will probably spend another few decades. I find two golf balls with the rising soil reaching to their midlines. Here's a strip of aluminum from a car windshield. The inevitable beer bottles and cans were probably tossed by teenagers tippling in the privacy of the bamboo. How long, I wonder, before the silica and aluminum rejoin the soil from whence they came? Glass I imagine is similar to obsidian and probably erodes at a comparable rate. Aluminum is more vulnerable to oxygen. Old beer cans seem to get flimsy after just a few years in the weeds.

Accompanying the human contributions to tomorrow's soil are some squirrelly donations. A jab of my shovel turns up a ghostly spike emerging from a black knob. Oops. My oak tree has for years been trying to produce an heir, and just when a forgetful squirrel gave it a break, here comes the spade. I poke the punky knob back into the dirt with low hopes. I presume squirrels are also responsible for the clusters of Norway maple sprouts striving for light under the vines. It looks as though the seeds were buried by the dozen. I know these sprouts would love to take the long route to soil, first sucking up every molecule of moisture and nourishment from my yard, starving out my berries, then thriving for a century or two before decaying. But—hark! I hear the squeak of hungry fungi. I bash the seedlings with the shovel and feed the mushrooms.

Every day I check on my beribboned articles. It's not heart-stopping drama. Over a few weeks, the leaf assumes a boat shape. It softens and darkens at the bottom, where it contacts the litter. An area the diameter of a pea is skeletonized. The apple also goes slowly. Even before it dropped from the branch, insects had punched holes in it, perhaps to deposit eggs. Once it fell, these punctures invited other diners. One day, I find ants converting a hole to a cave. Another day when I lift the apple, a beetle scuttles down into the earth. Gradually, the holes widen. Brown spots expand. Each puncture lets out some water, which comprises most of an apple. The apple shrinks. Tunnels now connect the apple to the depths. Miners commute up and down, stirring apple into soil. I, too, am ready to tunnel down and see what I can see.

I need to dig holes anyway, to plant the posts that will corral my tamed berries. The first few bites of the shovel lift a chocolate layer of earth. The white roots of lawn plants weave a mesh through it. After a moment of stillness, the cliffs of my shallow hole break into motion. Ants emerge to swirl across the opened earth like birds. My next slice gets under the chocolate, and into a crumbly cocoa of dirty sand. Rocks the size of arrowheads jut from the wall. At nine inches I spot brown glass. I shovel this layer into a screening box and shake it. When the soil is gone, Eurojunk remains jumping on the wire: a charred bean half the size of beans I know; clamshell fragments; a bit of glazed pottery. Back in the hole, I've opened a layer that's home to a species of ant almost too small to see. They're beige specks that glide over the cliff.

At fifteen inches, only the occasional root protrudes from the wall. Here the cocoa meets an orange layer of sand. The rocks are bigger. Old tunnels of moles and worms are recorded as tubes of dark soil that dive into the orange layer. This reminds me of the soil-tilling that my Cheeky chipmunk is undertaking. His tunnel system is many yards long, linking any number of escape holes. And as he burrows, he aerates and mixes the soil, and on his silken pelt he carries helpful fungi from one part of the lawn to another. Cheeky's homemaking maintains healthy soil.

At eighteen inches, the rocks consort too thickly to shovel. When I lift them out, they fall apart in my hands like loaves of sliced bread. Frost has split them where they lay. The only root I encounter is that of Asiatic bittersweet, a tree-strangling, invasive vine, which I savage on principle. At twenty-four inches, I encounter the maroon cone of a night crawler's head. Not wanting to injure him, I tap his noggin. He's either too torpid or too stubborn to back down. I cover him carefully before setting my post.

As I screen the dirt from the lower layers, I see why the people who once farmed my yard grew apples. It's all rocks. Plowing this stuff would have been like raking billiard balls. But the fact that six inches of soil lie atop the Eurojunk gives me pause. It seems kind of deep, considering that soil generally develops at a snail's pace. The global average rate of soil formation is about an inch every 450 years. Perhaps someone leveled the land when the houses were built, covering the garbage. Or maybe an old compost pile sat here, or a leaf pile. The possibilities are beyond number, says Dr. Samantha Langley-Turnbaugh, an environmental scientist at the University of Southern Maine.

"The traditional soil-forming factors fly out the window," she explains when she visits my yard. "Especially in the dense urban areas, you find fill, fill, fill. It's totally unpredictable. And you smell things that make you think, 'Hmm. I wonder what that is.'"

The most obvious difference between town soils and farm or forest soils is that people tend to compact the soils they live on. Good soil is like good cake: It should be about half crumbs, held together by fungi, worm slime, and other glues. The other half should be pore spaces, through which air and water flow. But when I walk across my yard, I squash air out of the cake. The soil has some resilience, but after enough stomping, the crumbs are not going to bounce back into place. The pore spaces close permanently.

"Roots need water and oxygen just like people do," Samantha says. "Without them, they die." She has been sending students to take core samples of the soil of Deering Oaks Park in Portland. The venerable oaks are ailing after more than a century of farmers' markets, fairs, concerts, and strolling. As for the soil—Samantha grimaces. "A half-meter core is ideal. We can't get the probe in any deeper than a half inch. We've broken so many of those probes."

Driving and walking over damp soil is the best way to convert it from cake to fudge. It's not uncommon for much trodden soils to exhibit a density identical to that of a brick. Even driving past them can do damage. The vibration from traffic can break down soil too deep to be squashed by feet and the occasional lawn tractor. It's one more thing to feel bad about when I drive: I'm not just heating the planet and commissioning the sloppy harvest of gloppy oil, I'm also wrecking the crumb structure of soil.

❧

For my next digging site I choose a patch of dead grass where a dog has peed. I don't dig far. The roots are as dead as the tops. The handful of sandy dirt smells sour. I pop it under the microscope for a better view. The roots look dry and yellow like day-old macaroni. The dust around them blows in gusts of wind. It's so dark and uniform that I think it might be eggs. But when I boost the magnification, the dust resolves into nodules of soil. Glinting bits of quartz cling together with brown and rust-colored particles. Nothing else moves. Nitrogen, which in small doses is the elixir of many plants, has burned this soil.

Nitrogen is one of those things that, like water, can be your best pal or your fellest foe. Although it makes up seventy-eight percent of the air, it's a form that plants can't use. In days of yore, a hungry plant had to rely on soil bacteria to produce "reactive nitrogen." In the temperate regions of the world, nitrogen was in such short supply that it was the limiting factor on the total tonnage of green things that could grow. The industrial era changed that. In the past 150 years we've multiplied the world's budget of reactive nitrogen fifteenfold. The biggest source is synthetic fertilizer, which both rises off and runs off farms and lawns. Another one third is made by bacteria that cohabitate with "nitrogen fixer" crops, and by the burning of fossil fuels in cars and factories. Lesser sources include sewage, especially when the makers of the sewage (Atkins dieters take note) eat lots of protein. Cows are prodigious nitrogen emitters, too.

You'd think this flood of nitrogen into the soil would be an unmitigated blessing. But it's mitigated. For starters, not every plant likes nitrogen. So if you add it to a landscape, you tilt the competition for sunlight and minerals toward nitrogen lovers. Studies on plots of land where nitrogen is added demonstrate that a few nitrophiles thrive, shading the majority of plants into extinction.

Under natural conditions, nitrogen falling with the rain onto my lawn would be massaged by plants and bacteria into an acidic, reactive form. But pour too much nitrogen into the system, and the soil becomes too acidic. This offends many plants. It also speeds the loss of important minerals from the soil. In the nitrogen-acidified water, they dissolve and wash into the nearest water body.

When I visited Katalin Szlavecz in Baltimore I also stopped by the office of Rich Pouyat, who's up to his ears in the Baltimore Long Term Ecosystem Research project. Rich, and a cast of dozens, will keep tabs on ecosystem indicators over many years, to spot trends in urban nature. Rich has been looking at how urban lawns compare to farms and forests, in terms of how readily they leak nitrogen. Peculiar news: Converting farms to lawns can be a net improvement, when it comes to nitrogen pollution.

"Agricultural soil is, in fact, extremely leaky," says Rich. He's a brick of a guy, whose office is stacked with books and backpacking gear. "Forests suck up nitrogen very efficiently," he explains. "You end up with those clean mountain streams. But not on farms." The reason is unknown, but it could be that plowing disrupts the fungi that should be capturing nitrogen. It could be that overworked farm soils run low on the organic carbon that nitrogen-fixing bacteria require. Or it could be that, in addition to menacing forests, earthworms are also ruining farms.

Earthworms complicate the nitrogen picture. The environment created by earthworms encourages the profusion of a kind of bacteria that convert soil nitrogen to a water-soluble form. This kind of nitrogen quickly washes out of soil. The result is "leaky" soil. And leaking nitrogen is a bad thing. When it hits a water body, it feeds an algae bloom. The ensuing algae demise sucks oxygen out of water. Fish die. Or, if the high-nitrogen water washes around a coral reef, it nourishes algae that smother coral. Coral die. Fish that require coral for food and shelter die. An earthworm can cast a long shadow.

Whether it's earthworms, plowing, or bacteria that make farms so leaky, it's too early to say. But right now, the farm is making the lawn look good.

"The lawn falls in between forest and agricultural soils," Rich says. "It leaks. But not as much as we thought." This analysis is preliminary, and the conclusions could change. For the moment, however, in the category of polluting the water with nitrogen, the lawn can lord it over the farm. Well, except for those lawns that get overdosed with fertilizer. Or on which a dog pees.

The Baltimore work has shown that lawn soils also host more microbes than farms do. That's a good thing, especially because some microbes can break down pollutants that lurk in more urban soils. Lawns also boast more worms, and more pill bugs and sow bugs, all of whom turn old vegetation into soil. And in contrast to forests, the plants and microbes in towns and cities may be able to consume much of that leaking nitrogen that worms overencourage.

❧

The final hole I dig is in a shaded patch that's surviving the drought. Up comes a plug of tangled roots, and the smell of dirt. Under the microscope these roots look as moist as fresh-cooked spaghetti. Still, something's missing. Between the drought and the season, the soil is devoid of all the creepy-crawlies who should be squirming through it. Where are the larvae of Japanese beetles and rose chafers? Shouldn't they be gnawing on the grass roots? By now, Japanese beetle mothers should have buried dozens of eggs, and the young grubs should be killing grass like crazy. Perhaps my skunks have hunted down every one. In any case, the soil ought to be jittery with tiny mites, springtails, and other little earthworkers who travel the worm trails and pore spaces of soil. Nematodes, a family of tiny worms who prey on everything from roots to root-eating insects, should be insinuated among the clover roots. But it has been horribly dry.

Fungi must still be swarming through this sample, but they're too fine to see. I get enigmatic greetings from my soil fungi when mushrooms pop out of the lawn. Like the tips of icebergs, mushrooms are mere fruiting bodies that rise from a hidden organism. Some underground fungal colonies cover acres. The first mushrooms to surface in my yard outlined a more modest support system. A ring of oystery caps circled an apple stump, suggesting that the colony was dining on dead apple roots. But the occasional singleton who raises his head as the summer progresses is more intriguing.

101

Is the odd, greenish mushroom the sole fruit of a small colony? Or is he the achievement of an organism who underlies my entire neighborhood?

The bacteria in the soil are likewise beyond my ken. From what I've read, they're present in this clot of roots in the kind of numbers only astrophysicists comprehend. "Several billion" per square yard. Depending on your definition of several, that's a minimum of two million bacteria per square inch, or eight to ten million in my hand. How many am I crushing under my feet at this minute? How many each time I mow the lawn? I push the plug of soil back into its hole. My palms are dirty—more will die. The dimensions of my crime are astronomical.

My ribboned leaf and apple will endure until they're covered in snow. The apple will shrink to one third its original size by the end of September, then one fourth by December. It will turn corky and tough. Most of its moisture will have escaped into the air, or into the guts of worms, ants, and beetles. When I lift it to check its underside, ice crystals will have formed between apple and soil. More crystals will be rupturing cells inside the punky fruit.

The ribboned leaf, downed in its prime by squirrel teeth, will lose more of its bitter tannin than the leaves that tumble down naturally in early winter. By snowfall, someone will have munched two big holes in it, chewing right through the tough veins. It'll be about twenty percent recycled.

By next spring, the eaten leaf tissue, and the apple pulp, the dead roots and rotting worms, the sacrificed Norway maple sprouts and even a few molecules of beer can, all will be contributing to a new round of life.

6

THE FREEDOM LAWN

ON A STICKY morning when clots of sea fog are trailing across the yard, I look out to see three of my crows lying in the grass. Their wings are spread wide. Heaven help us, have they got West Nile? They shudder, in turns. Their feathers flutter against the ground. Then they bob their heads. They're not dying, they're bathing! I've seen them flap in wet grass before, but this morning there has been no rain. My subsequent investigation convinces me that they're medicating themselves with my lawn plants.

Plant medicines are well-known in the animal kingdom. Overpregnant African elephants eat the same plant that their human neighbors eat to induce labor. Grizzly bears rub their faces with a chewed-up root whose compounds may function as bug repellent. A Costa Rican monkey treats its fur with three of the same plants that the people around them use on their skin. For that matter, crows are known to engage in "anting," whereby they crush ants and rub them through their feathers, presumably exploiting the formic acid in one insect to drive away another. But I haven't noticed anthills in the part of the yard where my crows are fluttering. What else could inspire a crow to bathe? I think it's time to survey my lawn from front to back and see exactly what it's made of.

Already I've determined that this lawn is on the cutting edge of turf management. It's a delicately balanced mixture of whatever can grow in any particular spot and doesn't mind getting whacked by the mower every couple of weeks. The lawn-industry term for this landscaping scheme is Freedom Lawn, capitalized by its coiners. The freedom applies both to the plants and the people, the latter of whom needn't water, fertilize, pesticize, or otherwise interfere. Nurseries now sell Freedom Lawn seed mixes, for lawn owners who live in parts of the world that have been shortchanged in weeds. But even before I came across the official term, I thought of my yard as a Darwin Lawn. Whatever could survive the mower and the drought was welcome to stay.

Whatever you call it, it's a popular landscape choice around here. Maine has been slow to recognize the genius of turning a perfectly good vegetable-patch-with-garbage-dump into an outdoor shag rug. And if we must have a shag rug outdoors, by gorry, we ain't gonna manicure the blasted thing. So I and all my immediate neighbors have Freedom Lawns. Which is just to say that a crawl across my lawn is going to be more like rambling across a diverse continent than walking through a tree farm. My Freedom Lawn contains the miniature equivalents of a jungle and a savanna, a prairie and a pine forest.

It's eye-blurring hot today, though. I'll crawl across the lawn later. Right now I'll take my books under the umbrella on the deck and read about the history of backyard grass-farming. Everyone's feeling floppy today. A squirrel spread-eagles himself on the shaded grass. When the turf warms up or his belly stops cooling down, he drags himself forward to a fresh spot and flops again.

The American lawn, in a nutshell, is a sterilized version of an English pasture. The French invented the lawn proper, but their turf was just another element of color and texture to deploy in the geometry of a formal garden. In the 1700s, English landscape architects began to veer off the garden path. The French tradition was a bit too urban for the evolving English view. The English view was that one visited one's country House to *get* a view, the vista in the metropolis having been obliterated by grime. And so, in front of a House in the Country, the English lawn was born.

More accurately, the lawn was renamed. It was there already, in that the House was surrounded by something called a sheep pasture.

But the fencing that kept the sheep from running off was considered detrimental to the view. This problem was solved, and the ideal lawn invented, in 1690. That was the year the ha-ha was born. The ha-ha was a ditch dug at a special angle. From the perspective of a sheep, the ditch appeared to have a fence leaning against its far side. But from the House, from the perspective of the guy who paid the landscape architect's bill, the lawn now looked like an open pasture that rolled to the horizon. Ha-ha! This became the lingua franca of landscapes.

The lawn evolved under the hand of Lancelot Brown, who so often referred to a landscape's capabilities that he became known as Capability Brown. In flat pastures, Capability saw the potential for smooth, rolling hills, and he hired laborers to reveal it. His famous lawns—some still grace English Houses today—romanticized the stuffing out of Nature. By Capability's time, English Nature had been deforested and overgrazed, but never mind that. Capability's pasture-lawns ebbed and flowed around instrusions of forest. He was always willing to eliminate a village or dismiss a formal row of old trees, if it facilitated the land's self-actualization.

So, the lawn became a sculptural landscape, inspired by artistic interpretations of the pasture-and-forest combo.

The lawn came to North America in roughly the same steps: first as pasture, then as artistic statement. When the English brought their livestock to the East Coast, they found the native grasses less nutritious than the grass back home. Plus, many native grasses were annuals, so if livestock ate the seed heads in the summer, the next spring's grass crop was sparse. By the mid-1600s, settlers were importing clover, "Kentucky bluegrass," and other European pasture seeds. The native grasses were overrun. In California the native bunchgrasses met a similar fate when the Spanish and their cows arrived in the 1700s. Mediterranean grasses now dominate that landscape. In the South, the African species guinea grass and Bermuda grass were running rampant by the early 1800s. And by the end of that century, Canada's bunchgrasses were yielding to imported timothy, alfalfa, clover, and other cow-fatteners.

As in England, the pasture evolved into a lawn when it rubbed up against a fancy house. George Washington had a ha-ha at Mount Vernon, along with a vast lawn mowed by deer. Thomas Jefferson, who admired Capability's work in England, transported the aesthetic to Monticello. He, too, had a ha-ha. Both these revolutionary gents relied on plants native to Mother England to green their swards. And both spent tons of money to exploit the capabilities of their estates.

Most people didn't have money to throw away digging ha-has and pasturing deer. A typical front yard was more likely to feature bare or weedy earth, with a border of flowers or shrubs. Particularly in the South, a dooryard of packed earth was a practical defense against snakes. The backyard was thoroughly utilitarian. It held a vegetable patch, chickens, berries, fruit trees, perhaps a cow, and certainly a rubbish heap. (It's interesting to me that the rubbish heap in my own backyard has remained in the same place through eighty-five years and numerous tenants. Recent owners piled old lumber, aluminum tent poles, and plastic milk crates in the same spot that early owners tossed their busted crockery, cans, and glass. Guided by who-knows-what instinct, I located my compost bin in there, too.)

More than a century passed before the average person could be badgered into beautifying the front yard à la Jefferson and Washington. It did take some badgering. Lawning was hard. The grasses people were urged to cultivate evolved in the rainy, cool climate of England. North Americans who didn't live in a rainy, cool area were hard-pressed to keep the stuff alive. They got iffy advice. One author directed people to sprinkle a failing lawn with salt and plaster, according to Virginia Scott Jenkins, author of *The Lawn: A History of an American Obsession*. And they had no lawn mowers.

But when people dispersed from cities to the new suburbs a century ago, they capitulated. In the suburbs, a man's home was his castle. And these suburban and village castles, the arbiters of fashion urged, ought to be miniature versions of Monticello and Mount Vernon. In the 1920s and '30s, the pro-lawn propaganda poured forth from garden clubs that were popping up like daisies on a dunghill. And these lawn enthusiasts had something to offer the beleaguered homeowner.

The seeds had improved, at no small expense to the taxpayer. The Agriculture Department, in cooperation with the U.S. Golf Association, had tried to find a grass for every climate and condition. The Feds had given some thought to fertilizer, too. The era of spreading manure (not to mention plaster) on the lawn was fading. The lawn mower had likewise matured. The first American brands, introduced in the late 1800s, probably compared favorably to the sheep, the scythe, and the horse-drawn mower, but they were nonetheless weighty, unwieldy, and unpopular. By the turn of the century, Sears, Roebuck & Co. offered four models at a modest $2.85, according to *The Lawn*. (A newfangled bicycle was going for about $12.) Furthermore, rubber hoses and sprinklers had been invented. Lawn rollers, fertilizer spreaders, and other aids were appearing, too. The lawn was putting down roots.

So, to recap once more: Immigrants brought European pasture plants to North America to feed their livestock. Wealthy Americans used those plants to imitate the lawns of wealthy Englishmen. The average homeowner imitated this imitation only when it became cheap and (sort of) easy to do so. Then we imitated with a vengeance. The lawn became de rigueur not only in North America's moist climates, but even in blazing deserts where a regimen of heroic measures are necessary to keep grass alive.

There were holdouts. There are still. New Englanders have a powerful immunity to fashion epidemics. Like rural people everywhere, I suppose, we stuck to our pragmatic landscape designs, even when we migrated into the suburbs. To this very day in my almost urban neighborhood, the occasional yard still functions as a site for lobster-trap storage, and/or parts-vehicles, and/or a vegetable garden. And I doubt I'm the only person on the block who flips my floor sweepings under the shrubbery or heaves rotten fruit out for the vermin, just as our ancestors threw spoiled beans and broken plates out the back door. And as I mentioned, most of us balked at striving for the current fashion, which dictates that grass alone belongs in a lawn.

Yes, it was not enough for a lawn to be rolling and green. As early as 1897, a USDA scientist was promoting lawns grown from a single grass species, and from which any intruding invader should be plucked, Jenkins writes in *The Lawn*. Clover held its own for a few more decades, but the writing was on the wall. The lawn would become a grass farm, with all the weeding, watering, and fertilizing that any farm crop entails.

Monocultures are sickly, that's a scientific fact. While it's true that some plants fight each other, it's also true that a high diversity of plants protects everybody. Together, the plants pool their talents. Each plant probably repels a few harmful insects. Each probably attracts a beneficial insect or two. The wastes of one species feed the roots of the next. The bold produce shade for the shy. They also dilute themselves, making it tougher for diseases to roll from clover to clover, or ryegrass to ryegrass. This has been proven in experiments: A plot of grassland hosting many species produces a lot more total greenery than a plot with fewer species. Even the most carefully tended plot of a single species can't compete.

And so grass-farming remains a struggle, even with all the chemicals. Replacing a mixture of robust plants with one or two species of grass is something like flattening a rain forest to plant corn. It can be done, but the law of the jungle will be against you every step of the way. Leaf-cutter ants will file in to snip up your corn. You'll blast them with pesticide. This will kill some of the beneficial insects you hoped would defend your corn. The soil, deprived of dead leaves, monkey manure, and ant tunnels, will soon wear out. The next time you buy pesticide, you'll buy fertilizer, too. The next year, you'll buy it again. And again. You're addicted.

The fertilizer pushers are happy, though. In the United States, we spend more money growing grass than any other crop, according to the authors of *Redesigning the American Lawn*. Perhaps this is to be expected, since home lawns cover more acres of the United States than any other crop. (Turfed roadsides, cemeteries, golf courses, and the like push the total higher.) As a nation, the United States spends upward of forty-five billion dollars a year on turf care. That's nearly half the gross national product of Ireland, spent on grass. And the dollars don't account for hours we spend babying the finicky plants. Or the polluting gasoline we pour into the lawn mower. Or the polluting chemicals we pour on the grass. The evils of weed killers and insecticides are real, and numerous, and frankly, so well documented that I find it tiresome to repeat the bad news.

Everybody knows our cultural fascination with grass-farming flits beyond the bounds of logic. But we do it anyway. We each have our reasons, and some of those reasons are even rational. My own reason is currently undergoing a major rethink. When I began studying my yard, my knee-jerk opinion was that the lawn was a waste of space that could be better spent on shrubs to shelter wildlife. Then I got to know my insects and realized that they feed the birds and mammals I'm so fond of. And insects, I discovered, think of my lawn as their beautiful prairie. So I was pleased to learn that the Freedom Lawn I currently maintain is a giant step toward sustainable lawn culture.

The Freedom Lawn was invented, so to speak, by three scholars at Yale's School of Forestry and Environmental Studies, who argue the merits of weeds in *Redesigning the American Lawn*. Their description fits my patchwork lawn perfectly. Who knew I was a trendsetter in landscape design? Poking around the Internet, I find that the Freedom Lawn is becoming more and more common. Lawn owners who lack the necessary weeds to start a Freedom Lawn are ordering "eco-turf" mixes of stuff like clovers, wild daisies, chamomile, yarrow, rye, fescue, and baby blue eyes. Some mixes combine flowering greens, while others feature herbs. Chamomile, one source advertises, perfumes the air with the scent of pineapple whenever you step on it. Some mixes are meant to be mowed only once a month.

The Freedom Lawn is becoming a movement. In Milford, Connecticut, residents now compete for an annual Freedom Lawn prize. The Garden Club of America, citing cancer rates, pollution, and water shortages, is promoting something called the New American Lawn, which sounds exactly like a Freedom Lawn. "Remember that Mother Nature abhors a monoculture!" the

109

Garden Club brochure exhorts. It goes on to urge a cold-turkey approach to pesticides, and patience while the soil recovers from systematic poisoning.

Even those who cannot learn to love dandelions can participate in the Freedom Lawn revolution. My friend Mark, an ardent environmentalist, cannot abide dandelions in his city lawn. But with two kids rolling in the grass, he can't abide toxic herbicides, either. "Corn gluten meal," he tells me, with a grass farmer's zeal gleaming in his eye. And from what I read, it sounds natural to the point of being edible. If you spread it in the spring before your clover, dandelions, and crabgrass germinate, it messes up their roots. I have so little grass that if I corn-glutened my clover and dandelions, I'd have no lawn left. But to each his own.

In my case, the Freedom Lawn movement is really a lack of movement. Before I was enlightened, I did feel a pinch of guilt whenever I skipped a week of mowing and things got flowery. Now I don't mow more than twice a month. And I'm going to let a patch go completely feral, to see what happens.

That, then, is the complete history of the American lawn: We've gone from pasture, to grass farm, and now back to pasture again.

Anyone curious about how weeds might suit their house can check out the Freedom Lawn at the White House. On a field trip far from my own two tenths of an acre, I found that the nation's capital has gone feral. Clover and dandelions, Canadian dwarf cinquefoil, and white-flowered chickweed twinkle up at me and the tourists who are peeking through the wrought-iron fence. I even see mushrooms. Biodiversity is running rampant. This is not what I expected.

I'm showing myself around the capital while I wait to meet the fellow responsible for all the weeds. I'm poking under leaves, counting ants, and squinting into bushes to spot birds. Here's a funny thing about urban ecology. Whenever I stop to inspect my surroundings, people gather. They follow my stare, and they see things, too. Evidently, they see things that are remarkable to them. "Look, Bob, it's a bird!" says wife to husband. "Hey, it's eating something!"

When I stop at the wrought-iron fence on the north lawn of the White House, my gaze draws the attention of a squirrel. The squirrel hops over and sniffs me for handouts. He's disappointed, but, lo! A small boy spots him and guides his family over. The squirrel scans the family for food. Now the squirrel is out in the open, and more people stop to watch him. He hops onto the sidewalk. Two men with Indian accents crouch, and one holds out a banana. The squirrel rises to clasp the fruit and gobbles. Seven people gather. Four take pictures of the man feeding the squirrel. A tall man in mirrored sunglasses, talking on a cell phone, stops to relay a play-by-play of the event into the telephone. More people slow down to watch. Strangers talk together about the squirrel. Until the approach of yelling Iranian protesters scares the rodent, a bubble of squirrel love claims the sidewalk.

The lawn to which the squirrel returns is begloomed with trees. Someone has raked the day's fallen leaves and twigs into a small pile. Half a dozen squirrels, one of them missing half a tail, nudge through the grass and weeds. That's a lot of squirrels in a small space. I'm reminded of a watershed moment in White House lawn history. Nixon was in office, and the rat population was exploding. The Interior Department distributed rat poison, but this incensed an unexpected constituency. Squirrel lovers rose up in outrage! Nixon's press secretary, confronted by a journalist with the possibility that the wrong rodent could be poisoned, considered the problem and made a policy announcement: "We're against the rats, we're for the squirrels."

This protectionist position persisted. Ronald Reagan was passionately for squirrels. He brought them bags of acorns from Camp David. Other visitors to capital parks fed them, too. If your tail was bushy, life was good. The nearby Lafayette Park, scientists announced in 1981, hosted the greatest number of squirrels per millimeter of any place in the cosmos. Although a gentle antifeeding campaign has trimmed the population, squirrels are still doing just ducky. Here on the back lawn of the White House, I see the first geriatric squirrel of my career, a beast so coddled it has grown gray and creaky.

111

Circling to the front of the White House, I spot a large raptor, perhaps a red-tailed hawk, high over the Mall. That has to be good news. It suggests that there's enough Nature in town to feed a family of large predators. You wouldn't know it from where I'm standing. A few songbirds hide in the trees, but I've seen only a single crow, and a handful of pigeons.

I proceed to the Smithsonian Castle, where Jim Sherald, pacing the stone-dust path, is easy to spot. He's tall, and he's the only man on the mile-long Mall who's wearing shiny, black shoes. Everyone else is in jogging shoes, football-tossing shoes, Frisbee-catching shoes, pram-pushing shoes, or ten-museums-in-a-day shoes. It's nearly raining, but not quite. A stone's throw down the Mall, planners of an upcoming festival have dumped truckloads of dirt next to their tent, killing many square yards of the nation's grass.

"It looks like they're creating a new landscape," says the man who strives to maintain the original landscape. "A landscape with hills."

Sherald, chief of natural resources and science for the Park Service's capital region, has done a lot of evolving in his job. Years ago, he felt it might be best if people kept their deadly little feet entirely off the Mall's struggling grass. Now he can shrug off the invasion of entire hills. And back when he started at the Park Service, the agency observed a regular schedule of fogging the Nation's Front Lawn with pesticides and herbicides. Now the Mall is weedy. A dandelion spreads its leaves at the end of Jim's shiny shoe.

Beneath the turf is a soil whose troubles are legion. Never mind today's imposition of mini-mountains. This soil has supported Million Man Marches. It has shuddered beneath horse races, and a railway line. Civil War encampments have come and gone.

"The soils are compacted to the point of concrete," Jim says. "Literally. And they're not native soils to start with." The native soil lies about twenty feet down. It's swamp muck. Filling began in the mid-1800s. The twenty feet of added earth is consistent only in its randomness. "You sample here, and then go a few feet away, and the two are drastically different," Jim says. A 1986 study of the Mall described in precise, scientific language a soil composed of building rubble and random loads of dirt. That even a dandelion can grow here seems noteworthy. I poke it with my toe. "Is that an abomination in your eyes?" I ask.

Jim looks down. "It used to be." But, he explains in a faded Southern accent, the whole Park Service has evolved. "There was a time when we had a treatment schedule. We treated just about everything. But the age of going up and down this Mall with a sprayer—hmmm." He casts a theatrical eye at the masses of joggers, Frisbee-ers, baby strollers, and tourists busily breathing air. "We don't do that anymore."

The chemical tide turned in the 1970s, with an increase in public awareness of the dangers. The Park Service questioned its own dependence on pesticides. "We started backing things out, and the world didn't fall apart," Jim chuckles. "In fact, the other day I was checking cherry trees for peach-scale bugs, and I couldn't find any. I think when we were spraying, we were killing a lot of beneficial insects that now keep things under control." So perhaps less spraying begets even less spraying.

The six hundred elm trees that edge the Mall require some medical care to hold off Dutch elm disease. But they're medicated only when they show signs of illness. And their chem-free green crowns may represent the biodiversity hot spot on the Mall, Jim says. "We've done sticky-card surveys in the canopy to gauge the insect activity. We found a surprisingly large diversity of life in the crowns. That was a revelation to me. This Mall is supporting far more diversity than you'd suspect. But it's not until you start looking that you see it."

As for the diversity of larger creatures, Jim inclines his head in the direction of a European starling prospecting in the grass. "You're looking at it." Crows, he says, used to be more numerous before the West Nile virus came. The occasional deer wanders onto the Mall, causing migraines for management. I, in a few hours of rambling, spotted mockingbirds and grackles and a few other birds, but that's about all.

However, the usual coterie of larger critters is easing closer and closer to town. Black bears, coyotes, and Canada geese are making pests of themselves nearby, and beavers once again have their eye on the cherry trees ringing the Tidal Basin. The day a coyote slips through the wrought-iron fence to nip the tail of a White House squirrel might be the day we're due for an updated policy on whom we're for, and whom we're against.

In all my capital perambulations, the only perfect lawn I can find lies in front of the National Sculpture Gallery. Not a weed raises its head. Each grass blade is as wide as its neighbor, and each is snipped off square. Still, something's going right. Three grackle youngsters beg and flutter in the grass, which means that insects are plentiful enough to grow birds. And a pair of mocking- birds are sitting on a red Calder sculpture. They pinwheel toward the grackle parents each time one comes to feed the children. They, too, probably have youngsters nearby.

It's all pleasantly surprising. I came here expecting to find the Nation's Front Lawn sterilized and embalmed in pesticides, herbi-cides, and fertilizers. Instead, I find it leading the way out of chem-ical dependency.

Cheeky finds me on the deck and explores me from toe to head, looking for seeds. I fetch him a cupful, and he shuttles them away. The panting crows ignore him when he pops out of the ground and darts into the shrubs, keeping cover as he returns to the deck. Cheeky has grown a size since I met him, and his tail is bushy. He carries more than thirty-eight seeds per cheek now, I think.

When the heat of midday finally passes and Cheeky has retreated to his tunnels, I put my nose to the ground and start my crawl across the lawn. I settle first in the bottleneck between the chokecherry tree and the lilac hedge. It's clover city. The clover leaves overlap, blotting every photon of sunlight. Beneath this closed canopy nary a soul stirs. The ground is warm, and hardened by drought. In the square foot I've chosen, not a stem of grass or yarrow, dandelion or sorrel, breaks the pattern of round leaves and white flowers. Wait, what's this fuzzy character? I open my wildflower guide. The interloper has canoe-shaped leaves with long hairs on them, and skinny flower stalks, atop which strange, hairy buds wave. Mouse-ear hawkweed. Dear me. What I have on my hands is a floral face-off. These two plants do not care for each other, and they're duking it out.

Despite the benefits of diversity, a plant won't let any old weed sidle into its territory. Clover (an Old English name for an old English plant) likes to live with clover. Hawkweed (the old English thought hawks ate it to maintain their eyesight) prefers the company of hawkweed. So, just as pines acidify the soil on which they sprout, so do hawkweeds. At least that's how it looks to scientists early in their inquiry. The soil directly below a patch of hawkweed is more acidic than the surrounding soil. The acid, in turn, boosts the level of soluble aluminum. Aluminum is toxic to clover, and to many other plants, insects, and microbes. And that's why scientists study hawkweed: With its pioneering spirit, it's elbowing the forage grasses out of valuable pastureland.

But my clover's not ready to roll over. Every plant excretes chemicals into the soil around its roots, and those chemicals encourage a particular suite of microbes. The activity of the microbes alters the soil to the plant's liking. Clover also nourishes bacteria inside its roots, and in exchange those bacteria convent atmospheric nitrogen to fertilizer for their host. The clover army is well supplied, and standing its ground.

Plant warfare is as old as the hills. In a world with limited soil surface, it's eat or be eaten, shade or be shaded. The Norway maple spreads a mesh of roots just below the surface, to catch water before it trickles down. The dandelion counters by opening a rosette of leaves that catch water before it hits the ground, funneling it toward a taproot. And here at the entrance to my lawn, clover and hawkweed spar with each other underground.

From the clover patch I crawl into a savanna. This spot is less shaded, and more parched. About eighty percent of it is covered with grass, and just one species. Mixed in is a smattering of yellow wood sorrel, one hawkweed, and a few of what I think of as "ten-petal."

(It has teensy white flowers. I thought bird books were a headache. Flower books are a cluster headache.) In this semiarid ecosystem, I can look down through the vegetation to the pale dirt. I see a pile of castings, the digested soil expelled from a worm's back end. I spy a bird turd. Ants of two species roam across the savanna. A dandelion parachute has tangled in the grass, and the seed is working its way toward the dirt.

The grass blades are ragged at the top. Dry fibers emerge like bones from green flesh. From a botanist's viewpoint, trimming vegetation with a mower blade must evoke the era of performing surgery with a hammer and saw. The leaves are literally clubbed in two. Oddly enough, this works to the advantage of the planet, if not the plant.

Turf mavens once thought mutilated grass like mine sucked up far more water than grass cut with a surgically sharp mower blade. The torn flesh, they believed, gushed water. But a mowing trial by University of Nebraska grass scientists reaches a Freedom Lawn–friendly conclusion. They find that grass bashed with a dull blade requires one-third less water than clean-snipped grass. Their best guess is that abused grass grows more slowly.

But wait. Nothing's clear-cut with energy and ecology. The scientists argue that we ought to keep the blades sharp anyway, because it reduces the mower's fuel use by twenty-two percent. Maybe the question is which resource we want to save the most. If it's energy, then we should be encouraging grass to leak more water, not less. Grass, like most plants, cools itself, and the air around it, by releasing moisture. Measuring the effectiveness of this trick is as easy as stepping barefoot from the lawn to the street on a hot day. A thirty-degree difference isn't uncommon. The contrast is strongest after sunset, when the grass cools quickly but the street throbs late into the night. All this cooling around the house reduces the amount of air-conditioning we need indoors. Grass is even better than trees at the evaporative-cooling business. (Trees win overall because their shade prevents sunlight from soaking the city in the first place.) So to save cooling energy, the more water grass consumes after a good haircut, the better.

Furthermore, if I mangle my grass I may damage its ability to make clean air. Like other plants, grass inhales CO_2 and exhales oxygen. Slow-growing, sickly grass doesn't do that as quickly. In truth, how effectively even healthy grass produces oxygen is a contentious subject. The Turf Resource Center, an industry group, says a lawn fifty feet square produces enough oxygen to sustain a family of four. Impressive! Baloney, argue the authors of *Redesigning the American Lawn*. The lawn consumes oxygen in any number of ways. Bacteria that digest grass clippings gulp oxygen, as does the gas-burning lawn mower. Manufacturing and shipping fertilizer burns more, as does pumping water. The average lawn, they say, is an oxygen and energy hog.

Debatable as well is how well lawns purify water. Experiments at a well-managed golf course have shown that polluted water running off a highway can benefit from flowing through clean, healthy turf. The grass slows the water enough that it sinks in. The aerated soil holds it. Microorganisms in the soil break down some of the pollutants. The water that comes out the other side is much improved. But if a lawn is not well managed, it becomes a source, not a sink, for pollutants. Water isn't going to benefit from flowing across a lawn freshly sprinkled with fertilizer or grub-killer. It's not going to improve itself by gathering dog manure or herbicides from a backyard. If a lawn is compacted or sloped, little water will sink in at all.

The same dichotomy afflicts a lawn's air-filtering potential. A lawn can trap lots of blowing dust and pollen. When I pluck a tuft of grass and clover to put under the microscope, I see that for myself. At high magnification, the leaves are spotted with hunks of quartz and flecks of mica, scraps of dead vegetation and yellow blobs of pollen. But whether a neighbor's lawn is a friend of clean air when she's blazing at dandelions with a bottle of herbicide—I doubt it. Whether my lawn represents a net gain for air quality when the low-performance engine of my lawn mower is burping soot—not likely.

Maybe the lawn's least debatable benefit is its recreational value. In my travels and interviews, I find that even those scientists most familiar with the evils of the American lawn tend to have one. "My boys like to play football," says one guy. Another says with a shrug, "I have to kick the soccer ball around with my kids, so I have to have grass. Then I have to put nitrogen on my lawn because it keeps the weeds out. If my lawn looks bad, my neighbors are going to be unhappy." And I admit I'd miss my own swath of the outdoor carpet. Sometimes I think I could live with no lawn at all, just paths between native shrubs. But just today, with a visiting dog, I sprinted barefoot around the yard. The Freedom Lawn was cool and sticky, and a shade of green that always reminds me of the Emerald City. It smelled green, and warm, and exactly like summer.

The most surprising ecosystem service I discover in my lawn is its ability to combat global warming. Remember the notion that planting a tree will soak up some of the carbon dioxide we release to the atmosphere by burning fossil fuels? Well, there's just a whiff of evidence that planting the right kind of lawn could remove even more carbon from the air, acre for acre. Studies of the amount of carbon stored in the soil beneath various eco- systems—farm fields, grassland, pasture, forests, and shrublands—usually reach the same conclusion: The pastures and grasslands sock away the most carbon. Letting woody shrubs take over a field is a carbon loser. Conversely, letting grassland take over a farm field is a big winner. Now, *grassland* and *pasture* aren't exactly synonymous with *lawn*. They're especially nonsynonymous with an all-grass lawn, because grasslands and pastures tend to support a diversity of plants, and their sods are deep and porous. But I'm encouraged that my Freedom Lawn might be related to this carbon-storing clan of landscapes.

Scholars argue over the role of fertilizer in this dynamic. Some say fertilizing the grassland increases its ability to store carbon; others say the manufacture and shipping of the fertilizer itself puffs more carbon into the air than it's worth. I don't fertilize, but the nagging question remains as to whether my Freedom Lawn stores enough carbon to make up for what's emitted by my lawn mower. I have low hopes. But it's good to know that my yard isn't a complete waste when it comes to climate change.

Enough of the grassy savanna. My onward crawl brings me near the spot where the crows were flailing this morning. What I discover is a cool hillside, thriving in the shade of the Bamboo Wilderness. To find the exact spot where the crows bathed, I look for bent leaves. I find a squashed fall-dandelion stem, and a few blades of grass that have been mashed against their neighbors. Beyond that, the predominant feature is a carpet of gill-over-the-ground, a plant whose purple-green leaves bear scalloped edges. I look for anthills that the crows might have been raiding, but there are none.

Perhaps the crows were exploiting the fall dandelions. I've seen Big Fat Momma the woodchuck grasping the tall buds with one black paw, and bending them down to her mouth. Is there something special about this plant? I chew a bud myself. It's not as bitter as a normal dandelion. The leaves smell rather mild. Maybe the gill-over-the-ground has something to offer. I crush a leaf and inhale an explosion of odor. The fragrance is like a combination of mint and grass. It's sharp, almost medicinal. Hmmm. To determine the medicinal potential of my weeds, I'm going to need more books.

I retreat to the air-conditioned library to gather treatises and histories of herbal medicine. The dandelion does have medicinal uses, I read, but they're related to its diuretic properties. It's recommended for premenstrual syndrome, which makes me wonder if Big Fat Momma hasn't been feeling a little bloated and irritable. The gill-over-the-ground, though, reputedly cures anything that ails me, or a crow. This humble weed is said to possess the qualities of an antihistamine, an antibacterial, an antioxidant, a cancer preventative, an antispasmodic, an antiviral, an expector-ant, a stimulant of the immune system and a sedative of the mind, and a cure for hypochondria, not to mention monomania. But it is also supposed to be an anti-inflammatory. The last trait interests me most. Bruised leaves are prescribed for skin wounds. And birds do suffer skin irritation when they molt and new feathers push through

119

their skin. Because this gill-over-the-ground is easily bruised, I can imagine that flapping in it might release its cure-all chemicals.

Well, that's my theory: Crows use medicinal plants in addition to medicinal ants. I float this balloon on a wildlife listserve I subscribe to, asking if any real scientist has heard of crows using medicinal plants. The silence is deafening. But I remember the similar response I got to my orioles eating the apple buds. So, you heard it here first.

Reading on, I'm pulled back in time. Many of the weeds in my Freedom Lawn, I find, are European antiques. Settlers who couldn't imagine life without these herbs carried them to North America when they immigrated. The gill-over-the-ground, in addition to its medical abilities, provided the flavoring for beer until hops edged it out. One of its many names, tunhoof, means "tipple ivy." No wonder they brought it. People can withstand many privations, but teetotalism is a rare feature in human cultures.

My lawn is a museum of useful European herbs. Besides gill-over-the-ground, I have soapwort, which settlers used to condition new-woven cloth, put a foamy head on beer, and wash the dishes. This I've got to see. I pick a stalk of the plant, which is conquering my flower beds, and rub a few leaves between my hands with warm water. Foam! The foam is green, and my hands are left smelling of leaf. But it's mild, and it works.

Yellow-flowered celandine, another misery to keep out of the gardens, was deployed against warts, liver disease, skin disease, ulcers, and cancer, I gather. Modern science suggests that many of those uses were justified, and a modified version of a celandine chemical is now produced as an anticancer drug. When I pick the plant, an orange latex oozes from the stem. I dab it on the flower book, where it hardens in a shiny dot.

Having solved my crow-bathing mystery to my own satisfaction at least, I turn my attention to the edible weeds in the lawn. My sheep sorrel, alyssum, and sow thistle are all recommended by weed-eaters. The Freedom Lawn is organic, but I wash my harvest anyway. You never know if a squirrel has sat upon your greens. I also pick young raspberry leaves for tea, avoiding those obviously patronized by a slug, who left silver slime behind.

120

I carry my salad up to my office, to make notes as I dine. Cheeky follows me and vacuums up sunflower seeds from the cup on my desk. His seed-trafficking and my note-typing are on a collision course. Not only does he add punctuation as he crosses the keyboard, he also drops seeds between the keys. When I bang the keyboard to clear things out, he scrambles to safety behind the computer. It's so touching when a wild animal trusts you. It's correspondingly irritating when he suspects you of attempted murder when you sneeze, run the printer, or clean the keyboard. We resume our respective meals.

The sheep sorrel, which glitters like spinach, tastes like a green apple. I'd eat that for recreation. The hoary alyssum, on the other hand, is tough. It's a member of the mustard family, and it has a spicy bite. Maybe I should have cooked it. The sow thistle is a tall, spindly thing that looks like a cross between bolting lettuce and a dandelion. The arrowhead leaves have the delicate crunch of a young lettuce leaf, and the bitterness of an older lettuce. Hmm . . . now I can't find it in any of my books. It's not sow thistle, after all, the leaf is wrong. Nor is it wild lettuce. Whatever it is, it's tasty.

Of course the settlers didn't have to make do with their imported vegetables and weeds. They also had access to a new world of weeds here. Native yellow wood sorrel is one of my favorite nibbles. The cloverlike leaves have a bright, sour flavor that vanishes so quickly that you want to eat another . . . and another. My lawn is full of it. Sumac is another plentiful native in my yard. Euell Gibbons, the guru of weed-eating, recommended an Indian-inspired 'ade of crushed and strained berries. But the rose-chafer beetles ate all mine this year, leaving me to speculate. Instead I brew a tea from the black-raspberry leaves. It's flavorless. Perhaps I'm supposed to be using the root. Or perhaps I shouldn't be using it at all. In a volume called *Iroquois Medical Botany,* I find black raspberry prescribed for "when a man is hunting and his wife is fooling around." It's also indicated for diarrhea and whooping cough. I'm not bothered by any of these at the moment.

Native ragweed, which I have in abundance, the Iroquois used for skin infections and stings, stroke, and "cramps from picking berries." I must keep that in mind. My old friend ten-petal turns out to be a native chickweed, which the Iroquois employed against miscarriage. In fact, it's a challenge to find a native plant in my lawn that was *not* in the Iroquois medicine chest. They were even quick to exploit new plants the Europeans introduced. Plantain, the fibrous brute that rules my compacted driveway, was adopted by Iroquois to treat the following ills: burns, sores, cuts, bleeding, bruises, spider bites, fever, sprains, bladder problems, nervous breakdown in overworked women, infertility, and arthritis. The English brought it by accident.

~

In late summer my lawn mower breaks. My Freedom Lawn goes extraferal and turns into a glorious thing. I get attached to it and don't ever want to mow it again. The plants that have endured decapitation every two weeks now raise their heads and bloom. They're stunted enough that the resulting prairie is about six inches high. A bonsai Queen Anne's lace produces a dwarf doily. The glossy item whose leaf reminds me of marijuana throws open brilliant yellow flowers. Microdaisies bloom at just four inches. Hoary alyssum send up thin stems of white flowers like rocket trails. The sheep sorrel makes an effort with gritty-looking red flowers. Ragweed is preparing to propagate in miniature, shoving up two-inch spikes of nubby, green blooms. Clover forms white clouds. A native Saint-John's-wort unfolds yellow flowers. Even a couple of latecomer dandelions lift their tubes to blossom. The Freedom Lawn becomes a micromeadow.

I'm pleased, and my outdoor neighbors are, too. Cheeky has better cover when he bounds from my house to his home with a cheekload of seeds. From the deck, he surveys his path for predators, then he leaps into the meadow, leaving stems quaking behind him. Big Fat Momma, my woodchuck, lies down and noshes on whatever's in reach. She reaches out a black paw to behead a dandelion, then lowers her head to snip a leaf. When she's harvested the best bits, she heaves to her feet and waddles a few steps before plopping down again. I can't tell if she's eating the native Saint-John's-wort with her diuretic dandelions, and I find no scientific claim that the native *(Hypericum punctatum)* contains the same cheery chemical (hypericin) as the European version *(H. perforatum)*. But it makes me happy just watching Big Fat Momma graze through the herbs.

The whole scene pleases me. My plain-green lawn has turned into a tapestry. I'm amazed that all these flowers were able to wait out the tyranny of the lawn mower. But I suppose they wouldn't be living in my Freedom Lawn if they couldn't handle being whacked. As I get to know them, I learn that some propagate by sending underground runners (hawkweed), others bloom and produce seeds below mower height (white clover), and some, like the dandelions, grow their buds flat on the ground until the last minute, when they raise and open them. But I wonder if they're evolving even as they live here. I wonder if I and all the other Freedom Lawners aren't creating new strains of midget plants whose blooms never rise to meet the blade.

When the mower comes back from the shop, I sigh and decapitate the wildflowers. But I do spare a patch. The apple stump is hard to mow around anyway, and I cut it a wider berth. I'm going to have a meadow four feet in diameter.

The plants in this Tuft surprise me. The grass matures into three different species, all more than three feet high. One is the cattail-headed timothy I recognize from the hayfields of my youth. Another produces seeds in two herringbone rows. The third dangles its seeds from threads, then tips over on the lawn. Each grass turns brown at a different time. One day I look out to see that a goldfinch has forsaken the sack of thistle seed I offer. He's perched in the tall grass of the Tuft, feeding on grass seed.

The wildflowers go crazy. Creeping bellflower erects stalks of purple blossoms. Ten-petal flings itself into the upper echelon, propping itself on the grasses. In the middle is the prize: Two sumac sprouts, their caterpillar-green bark fuzzed in magenta velvet. Indians used sumac as a spice, I read, and to cure convulsions in children, bellyache in horses, and irregular menstruation.

Depending on the ailment, they would scrape the bark up or down the branch. They even smoked the dried leaves. Sumacs are difficult to plant and spread best underground, as these have. I adore them.

Nosing into the Tuft, I see that others are finding what they want, too. A mouse, or perhaps Cheeky, has established a set of tunnels through the grass, all leading to a fresh hole excavated under the old stump. At the moment, the entry to this burrow is being used for a mating cave, by a cricket. The Tuft is turning into a microhabitat. I start planning a whole series of Tufts for next year's lawn.

My Tuft, though, could be in violation of South Portland law. Many towns have weed laws to deal with people like me. These ordinances generally forbid "noxious weeds" and stipulate a maximum height for turf. The usual rationale is that a long lawn gives shelter to snakes, rats, and man-eating tigers. But that's rarely what neighbors complain about. It's the way a wild lawn looks that grass farmers find irksome. It's the way weeds try to march out of the Freedom Lawn and into the manicured grass.

As always, I'm as hypocritical as the next girl. One of my neighbors, an absentee landlord, has let her backyard go to Japanese knotweed, commonly known as bamboo. This plant has world domination on its mind. In the few weeks that my mower was broken, the bamboo pioneers on my side of the fence shot up three feet. Left to its own devices, the Bamboo Wilderness would invade and shade out my mini-meadow. Big Fat Momma would find nothing to eat. The crows would take their medicine in some other yard. The goldfinches would go elsewhere for seeds and would nest elsewhere. The insects would come in swarms during the week that the bamboo blossomed, but would otherwise skedaddle. My yard would become a monoculture. And crawling from one end to the other would offer the allure of a ramble through a field of corn.

❧

Curious about the extremes of lawn culture, and the lengths to which some people will go to create a green carpet, I leave my yard again. This time I head for Arizona, a state designed to grow cacti and conifers, not thirsty grass plants. I'm not out of the Phoenix airport before someone is talking to me about water. The rental-car clerk asks what I'm doing in town, then shakes his head. "I'm from L.A.," he says. "I couldn't believe it when I saw the concave lawns here. People flood 'em four or five inches deep and let it stand overnight. I thought a water main was broken."

Something's broken, that's for sure. Phoenix receives seven or eight inches of rainfall a year. That's enough to grow cacti and decid-uous plants that spend much of the year looking dead. I get about six times as much precipitation at home. But Phenix is wall-to-wall green. From the window of an airplane, it looks like a patch of Merry Olde England, rolled out in the brown desert—Merry Olde England with palm trees. Up close and personal, some innovations become evident: Many homeowners have done some "xeriscaping," replacing grass with cacti and other drought-tolerant species. But even *more* up close and personal, maybe this is not an improvement.

Dr. Chris Martin studies the effect of the transition. I meet him at the Arizona State's Center for Environmental Studies, which squats in a treeless, 108-degree parking lot. Chris is a lanky guy whose wee, wry grin is a near-constant fixture. We climb into a truck and crank the AC.

"Before air-conditioning, no one gave a flip about xeriscaping," he says, pulling into traffic. "The landscape around a house had a function. The trees had to provide shade. A yard had to cycle water to cool the air. But now if your house is insulated to R-20 or R-30, shade doesn't matter. The landscape is now aesthetic."

He pulls to the curb in front of one of those sunken lawns the rental-car clerk mentioned. It's greener than my moist, Maine yard in springtime. This yard, Chris explains, inherited an agricultural water-allotment when farmland was divided into house lots. On a regular schedule, water flows from a canal into the yard. This yard sports fat trees, shrubs, and copious shade. Oranges have plonked into the street.

125

"These houses predate air-conditioning," Chris notes. "The trees were planted for shade. You'll also notice as we go along that there are lots of birds in these flood-irrigated yards." There are. It's cooler on this street, too. And less squinty-bright. Not in evidence, however, are the redirected rivers (the Verde, the Green, the Colorado) that keep it all going.

"Every single plant that you're looking at here is being irrigated—by flood, sprinklers, or drip irrigation," Chris says. "If it weren't, it would all be dead in a couple of months. The big trees might take a year. There's a lot of water going into this system." This is the wet end of the spectrum of Phoenix lawns. We drive to Chris's own house to see a middling example.

Chris's small front yard is xeriscaped, mainly with rocks. Around the edges of his large backyard are some desert plants—a green-skinned paloverde tree, a massive mesquite. A quail thrashes through the shrubbery. A vegetable garden withers in one corner. But the bulk of his acreage is in grass. "I have six kids," he says. "The five boys like to play football." So he waters.

Now onward to a real xeriscape. Or is it a real xeriscape? Chris stops the truck at a mansion in the hills. Crushed stone paves the lawn. An ocotillo cactus stretches slim arms toward the sky. The chubby barrel of a young saguaro rubs spines with a sprawling prickly pear. I think it looks terrific.

"It's a cartoon," Chris says. "This is the landscape we call Disney Desert: small turf patches, with irrigated desert plants. Because the fact is, no one wants to live in a desert. We've found that the longer a person lives here, the more they prefer green. Look: That's real desert." On the other side of the street, the houses end and Nature begins. It's scruffy and brown. The sunlight flattens it into two dimensions. It's deadly boring to contemplate.

"Okay," I say. "So if these people wanted a real desert, they could have creosote bushes and saguaros?"

"*One* saguaro," Chris corrects. "And not irrigated so it's obese."

Irrigated? But of course, says Chris. Though the Greek root *xeri* means "dry," the industry that now flourishes under that name specializes in the *illusion* of dryness. Perforated hoses snake beneath

126

most desert scapes, oozing water into the soil. The cacti stay fat, and the shrubs hold their leaves.

"Xeriscaping is not working at all, if you look at it from the standpoint of water conservation," Chris concludes. "If we did it well, we'd put the landscape-maintenance industry out of business. And most of the people in that industry are low-income people. Do you want to put those people out of work?" His ironic grin is cranked a little higher.

Furthermore, he reminds me, the flood-irrigated neighborhood where we began our tour was cool and shady, wasn't it? That shade keeps the entire city a little cooler. And those water-hog trees, grasses, and shrubs cycle water much faster than desert plants do, and that evaporation cools the city even more. So which would you have? More water use and less air-conditioner use? Or less water use and more AC? A water shortage, or fossil-fuel smoke in the air? Chris chuckles. Heh, heh. He has no easy answer. However, the university owns a plot of Phoenix houses and plans to landscape them three different ways to compare the effects. Not that Chris expects science to change anyone's mind about his or her lawn.

"I imagine," he says cheerfully, "that we'll all waltz happily along and then we'll have a crisis."

I say good-bye and head south through the desert—and emerald-green squares of irrigated corn—to Tucson. At the other end of the extreme-lawn spectrum, two brothers on a small urban lot are trying to live within the water budget that Nature allotted to southern Arizona. Even though they grow food around their bungalow, Brad and Rod Lancaster still use less than one fourth the water their neighbors consume. They could almost get by with the rain that falls on their soil and their roof, if they could bring themselves to chop down their sour-orange tree. But we all draw the line somewhere.

I have a hard time finding the house, because it's behind a hedge of mesquite, fruit trees, and sickly looking cacti. The gate, when I find it, is a sculpture welded from bike parts and tools, one of Rod's specialties. Over the hum of a Rod-made fan (the brothers are AC-free), Brad hears my knock and comes to the door. He kicks on sandals and slaps a hat over his blond curls. It's even hotter here than in Phoenix. But Brad lifts his freckled nose to the wind and looks surprised. "That smells like rain." It does. And it hasn't smelled like that for a long time. Tucson has been gasping from drought.

"This is the barest time in the garden," he apologizes. "We let everything go dormant, and the chickens take over from April to July."

Chickens. I spot a black one with a halo of chicks in the underbrush. "It's not legal," Brad confesses. "But we went to all our neighbors and asked them to give us a month. If the crowing didn't blend in with the dogs and birds in a mouth, we'd get rid of them. But no one has complained." Sharing the eggs with the neighbors probably softens the impact of the two roosters. Between eggs, cactus fruits, domesticated fruits, native seeds, and vegetables, the brothers produce between fifteen and twenty-five percent of their food on this one-eighth-acre lot. In addition, they produce some of their own power, with a solar oven, solar water heater, and a solar-electric panel. And they also produce water. Brad is writing a book on rainwater harvesting, and he's overflowing with statistics and sentences.

"You have to plant water before you plant plants," he says, quoting someone he met in Zimbabwe. He starts pointing out water traps, all around me. The yard is sculpted into basins and troughs, so that not one drop of rain rolls away. He pushes aside some tree branches and reveals a twelve-hundred-gallon water tank on a pile of rocks. It collects all the water that runs off the house's roof. We move toward the back of the yard and find an outdoor shower, curtained with dry palm fronds. Instead of a single drain, it has three. Brad extends one foot to shift a rubber plug from one drain to another. Before, or even during your shower, you can decide which fruit tree to send your bathwater to. Outside the yard, on the public right-of-way, the Lancaster brothers have been just as thorough. They've cut through the curb between their property and the street. Now when it rains, the runoff that usually rushes down a storm drain is diverted into holding ponds. I don't ask, but I imagine these guys even direct their pee to this or that corner of the yard. Water is water.

With water in place, the brothers planted their plants, large and small. Brad now consults for others who want to harvest rain and grow native plants. His underlying motivation seems to be a love for the desert surrounding Tucson. Here, and in clients' yards, he reintroduces native vegetation. He's extrafond of plants that produce food (the Sonoran Desert, he says, offers 375 such plants), or that have a medicinal use. He nibbles as he introduces me to his plants and plucks things for me to chew. "It's a living almanac," he says of his yard. "When I see my cholla blossoming, I know I've got to get out and harvest them from around the city. When I see my prickly pears fruiting, I know they're fruiting out in the desert."

"Is this a prickly pear?" I ask as we pass the beehives and the chicken coop. "And what's wrong with it?" The thing is wrinkled and droopy. Brad looks a little surprised. "That's what the plant does naturally in the dry season," he says. "We get enough rain in Tucson to take care of a native-plant yard like this. But people still water, because they don't like it when a plant doesn't look good. Even though that's the plant's natural adaptation." But aesthetics should be no impediment to converting Bermuda grass to prickly pears, Brad insists. "That's where gray water comes in. The wastewater from our lives is enough to grow a jungle." The Lancasters like to say they consume their dirty laundry water in the form of peaches. And sour oranges. Brad sighs. That tree does suck up water, but the leaves are so good in Thai food.

Outside the fence again I start to appreciate how wide this yard's circle of influence extends. (Rod, dark-haired and reserved, is no less zealous than his brother.) The brothers have wild-ified the unpaved sidewalk with more cacti and trees, making a shady path for pedestrains. Neighbors have begun to add native plants to their own yards. Birds are carrying native seeds from the Lancasters' yard to more distant yards. A few blocks west, the brothers led an effort to convert a parking lot to a community garden. Every year they organize a native-foods festival to introduce people to the Sonoran Desert's abundance. They've spear-headed tree-planting campaigns. Brad is currently prodding the city to let people cut their curbs the way he has, to capture street water. And so on.

I ask Brad what he would take as a sign of success, an indication that Tucson's yards were functioning as a desert ecosystem. He ponders, a solemn young man with the fate of the planet on his mind. "In the neighborhood, I'd like to see quail come back. That would tell me there's good cover to protect them from cats. And, you know, my parents used to see javelina, bobcats, mountain lions, mule deer, and white-tailed deer around their house. Those things belong here. We did actually have a javelina come through the neighborhood once. And since we planted our own yard, we've gotten cactus wrens, curve-billed thrashers, flycatchers—they eat our honeybees—geckos, lizards—the chickens eat them—horned toads—the chickens are a little hard on them, too—butterflies, sphinx moths . . . beneficial wasps . . ." He thinks a bit more.

"If I saw willows, cottonwoods, and Mexican elderberries starting to volunteer, then we've done it. They used to be all along the river and the flood plain. People would get lost in them. Right now, we're desertifying this area, and making a more arid ecosystem. But if those trees could make it, that would indicate that the water table was up to where it used to be. The same with native grasses."

The wind is picking up, carrying the scent of hot rain. Brad has a meeting to get to, and he climbs on a bicycle. I head off to the mall for a fresh tube of sunblock. As I park, thunder cracks. Golf balls of water splat on my windshield. As I cross the parking lot, waves of water sluice over the asphalt. Under the store's awning people shift excitedly. They look at their cars, then up at the lightning and water balls. "Wet T-shirt contest!" one woman shrieks, and sprints for her car. The others laugh and shuffle their feet. Another woman says, "We don't know what to do in the rain."

At the edge of the lot, the rushing water sloshes against an asphalt island planted with emerald turf and shrubs from Asia. The water bounces off the island and flows away down a drain.

FALL

7

BEFORE

MY NEIGHBORHOOD AND its beach are named after Captain Benjamin Willard. In a portrait, the man stands solemn-faced. His mustache is elaborately curled. His tights pouch at the knee. The fringe on his shorts tickles his thigh. A spotted dog, also solemn, sits like a hat, on Willard's head.

The captain's 1895 autobiography includes tales of sharks and swordfish he caught. It recounts shipwrecks, Civil War skirmishes, and other acts of derring-do. But the chapter in which Willard's gentle soul shines, the chapter most heavily illustrated with drawings and early photographs, is the one titled "My Trained Pets, the Coach Dog and the Cedar Bird." That chapter opens with a drawing of the big man lying on his back on the parlor floor, one leg extended toward the ceiling. Balancing on Willard's foot is, again, the dalmatian.

One can view the history of a place through any filter one chooses. Why not through dogs? They're the poster pups of the modern backyard. And prehistoric dire wolves, foxes, and other canines probably inhabited my yard before people did. Then, when the first people walked in twelve thousand years ago, a domesticated canine came with them. Most of the local canids fell into extinction. Most recently, Europeans sailed in four hundred years ago with a fresh batch of dog stock. Once again, most of the resident dogs faded in the face of today's dominant strains. Dog replaced dog replaced dog. One of the wildest chapters in this doggy history is unfolding right now, as coyotes and wolves soft-foot back toward my suburban lawn.

These days, the dominant canine sleeps indoors, and from among

133

the odors of shampoo and coffee, he can't detect the scent of his approaching cousins. Cousin coyote, whom wolves once kept penned in the West, has lately trotted as far east as Nova Scotia. Like crows, the coyotes are finding the suburbs a safe and agreeable habitat. Cousin fox, too, survives to hunt the suburbs. I recently watched one trot down a nearby street in daylight. Even wolves, extinct in Maine for a century, are padding down from Canada. One made it all the way to the coast before being snapped up by a trapper a few years ago. They're also moving toward the yards of Minnesota, Michigan, Montana, Idaho, and Washington. I'm fascinated by the possibility that wolves might decide they're just as comfortable among the bungalows as the coyotes and foxes are. It would be a curious circling back to the first meeting of tame dog and wild.

Every canid that ever crossed my yard traces his ancestry to a pack of weaselly critters who climbed in trees about sixty million years ago. This new branch of mammals, called miacids, wanted a meatier diet than their predecessors enjoyed. When the dinosaurs died, the carnivore job was suddenly available. Miacids evolved big brains, which they needed to sneak up on prey, and big teeth for ripping open the meat packages. When the group divided about fifty million years ago, one branch turned doggish, and the other turned cattish. The doggish branch split further. The ancestors of bears and seals, which have always reminded me of dogs, branched off, along with increasingly doglike animals. Some thirty million years ago, a German shepherd on dachshund legs may have scuttled through my yard. It looked like a dog.

At the time, my neighborhood was taking part in a global cooldown. A thousand miles north of my lawn, a ribbon of tundra was easing southward. In this final ice age, what did the protodogs hunt? Maine has no useful fossils from that time, thanks to glaciers plowing them into the sea. But judging from backyards in other parts of the nation, my yard was home to minihorses, macro-pigs, proto-camels, rhino also-rans, mammoths and mastodons, and perhaps even bear-size beavers and giant sloths.

The shepherd-dachshund charged about the neighborhood catching bizarre animals for many millions of years before his line petered out. Umpteen more pre-dog species continued to evolve toward dogness. The model that finally succeeded was a fox from the North American Southwest. From him descended today's assortment of wolves, foxes, jackals, coyotes, as well as wild and domestic dogs. Most of this speciating took place outside of North America. The protowolf loped away to Asia early in the game. There he evolved into the gray wolf. In Asia, people eventually tamed the gray wolf. Then the wolf, the dog, and people all journeyed from Asia eastward, toward my yard.

When Nature's canine experiment began, that original foxlike critter could have skittered from my backyard to Greenland and Europe. But the opportunity to hop continents was ending. The continents were crawling apart at about the speed my fingernails grow. Soon, North America would be connected to Asia and Europe only in the west, by the Bering land bridge. And that was prone to flooding as the sea level fluctuated. After the ice ages commenced two million years ago, a glacier often blocked our end of the bridge completely.

Some 160 million years prior, my yard had been down by modern Australia, mashed against northern Africa. Dinosaurs romped in Maine, although my yard was probably too alpine for their liking. When North America and North Africa initially collided to form Pangaea, they plowed together the sea sediments that had separated them. Mountains rose. The roots of those long-eroded peaks are still visible a couple of blocks downhill from my yard, at Willard Beach. The folded, crushed layers of old seabed today form ragged ledges. (Mine wasn't the only yard altered by its Pangaea experience, of course. At the Pangaea party, the backyards of the Southeast waltzed with the western Sahara, which put the final touch on the Appalachian Mountains. Our subsequent departure from Pangaea tore a chunk of Africa away, to become the backyards of Florida, southern Georgia, and coastal South Carolina. As North America backed away, it buckled the Pacific seafloor, giving a boost to the Rocky Mountains, which at the time formed the West Coast. The backyards of the current West Coast were manufactured as the Pacific plate continued to collide with North America.)

It wasn't until the time of the dino-die-off sixty-five million years ago that North America began to resemble its current self. An inland sea drained from the Great Plains of Canada and the United States. The widening Atlantic opened new arms, separating North America, Greenland, and Europe. After that, North America was often isolated.

It was practically yesterday that real wolves cantered into North America from Asia. Just a few hundred thousand years ago the ancestor of the gray wolf made his debut in my yard. (Well, he did if the ice didn't get there first. Glaciers have been a recurring problem in my yard, and one was on the way at that time.) What a feast of fantastic creatures awaited this new wolf! He, along with local saber-toothed cats, lions, coyotes, and the enormous dire wolf who evolved in the Americas, must have left a heap of strange bones in my yard. Imagine the wealth of fossils!

Then imagine them a hundred miles offshore where the glaciers dumped them, probably after pulverizing them to fossil powder.

The wolf and dire wolf, who bore the heavy jaws of a bone-cracking scavenger, may have leapt into my neighborhood between ice ages, just 125,000 years ago. The sheet of ice that has mashed my yard for most of the last two million years had shrunk toward the pole. At that moment, my particular lawn was beneath the Atlantic, which was swollen with the water of melted ice caps. The higher hills of the neighborhood were covered with forest. The wolf may have run down ancestors of today's rabbits and woodchucks, and even the forebears of my Cheeky chipmunk. The dire wolf, similar in size to the modern gray wolf, might have gnawed lion bones in the woods or scrounged dead fish from the shore, which would have been at the top of my street. Perhaps to cool off, he dog-paddled over my yard.

But the ice soon rolled back down. A snowbank a mile deep scraped over bedrock toward my lawn, plowing up soil and boulders. As the ice sheet expanded, the sea starved. It was donating snow to the glaciers, but the glaciers weren't generating rivers to refill the ocean. The ice swelled, and the Atlantic receded.

Long before the glacier arrived on my lawn, cold winds poured off the looming ice and chilled the ground. One year, the acorns from the neighborhood oak trees never warmed enough to germinate. A few decades later (the glacier's speed can only be guessed) frost crystals bit so deeply into the oak twigs that the oak cells were rent. The following spring, the insect who laid her eggs on an oak doomed her offspring to starvation: The trees would not leaf out. Maybe a few hundred years later, scrappy conifers were the only trees still able to wring a living from my cold yard. Tundra grasses now found it hospitable. One spring, the frost never came out of the ground.

As the tundra plants fled to Connecticut, the beetles and lemmings who ate those plants fled with them. The foxes who pounced on the lemmings followed their dinner. The dire wolf drifted south, too. My yard grew quiet, except for the swish of the wind over frozen ground. Then the ice came. Over in Portland, it crawled up one side of the peninsula and rushed down the other. The snout of the ice was filthy with plowed earth and accumulated dust. Its shell was brittle and cracked. The rolling ice-bank crossed Portland Harbor and rumbled up Willard Beach, wrenching out cubes of the bedrock and dragging them along. Up the hill it came. The ice plowed my yard to the bedrock and continued southward clear to the continent's edge 250 miles offshore. On the bedrock it left nothing but tracks.

I bundle up and walk to Willard Beach to track the glacier. The wind lately is feeling a bit glacial against the skin. At the ocean, I hear the unfamiliar voices of the Beach Crows, my clan's northern

137

neighbors. Like my crows, these are only around during daylight now. Each late afternoon, they join puddles and rivers of crows that flow from all directions toward Portland's Deering Oaks Park, which they've chosen for their winter rookery. The park trees fill with a thousand crows as the sun dims. The streetlights blink on. Mornings, my crows come home spattered in the white excrement of their bedfellows. Here at the beach, the local crows are pulling dead fish and crabs from the seaweed. Like my crows, they can tell when they're being watched, and they don't appreciate it. I keep my eyes on the rocks.

When the glacier came through in the last ice age, boulders and rocks embedded in the ice acted like giant sandpaper against the old, folded rocks. I don't have to look far to find their scratches on the exposed ledge. This outcrop is a worn-down fold of sedimentary rock dating to the Maine–Africa collision. The layers now stand on edge. Most stripes are the color of graphite, but ribbons of quartz intervene.

When I face south, the low autumn sunlight throws shadows into grooves on the rock. Most of them could be mistaken for marks left by a pebble stuck in the tread of a boot. But there are too many. And they're all scratching in the same direction. And some of them are a half-inch wide. That would be a big pebble. On hands and knees, I crawl along a shelf of rock, tracking individual pebbles, and the motion of the ice that pressed them into the ledge.

Psuedoscientist that I am, I'm worried that my "glacial striations" may have been made by pebbles in boots. I survey the whole ledge. Crawling on, I find striations on the corners of little cliffs where people couldn't walk. More scratches emerge from under soil. Okay, I'm convinced. But now I've found a small patch of folded rocks peeking out of the sod. Through it runs a vein of quartz tinted with iron that has oxidized. Garnet? I think so. A second quartz layer is exposed to the elements, and its softest minerals have dissolved out, leaving an open matrix of crystals. Gorgeous. I crawl on. Every crystal tells a tale. Continents rove. Glaciers rumble. Dire wolves flee. The sun dims. The crows rise

138

to the trees, then rise again and head for Deering Oaks. I rise, too, and run for home.

❦

Every backyard in North America shivered when the ice came south. The white wall crunched over New England and New York. It overwhelmed the Great Lakes and grated across Ohio. From Manitoba it rolled over the Dakotas and Minnesota. It mashed Bismarck and Seattle. When the ice stalled eighteen thousand years ago, it was flanked by a ribbon of tundra that ran through the Midwest and Pennsylvania. South of that, most of the continent's backyards grew a winter coat of frost-worthy pine and spruce.

Just how the camels and sloths and dire wolves felt about the chill is a subject of debate. They all disappeared as the ice finally waned. But whether their dietary needs couldn't keep up with the changing plants, or whether a new predator and his dog ate them all, only time and research can tell. Whatever the case, Man and Dog did walk into North America near the end of the last ice age, and things would never be the same.

Perhaps that's not saying much, considering how dramatically the glaciers were rearranging things. Even as people from Asia fanned across North America, ice continued to renovate my yard. When the ice advanced, it plowed the bedrock clean. Then as it melted, it dropped all the boulders, gravel, and sand it had carried. To see the "before" and "after," I unfold two geology maps.

On the Bedrock map of Maine, I can make out the roots of the old Alps, now hidden under my backyard soil. The layered rocks were folded into ridges, then erosion and glaciers grated off the ridge tops. What's left is a corduroy of different-aged layers tilted on edge. To see what the melting glacier dropped atop my bedrock, I turn to the Surface Geology map. Gray bedrock pokes through the glacial junk in only a few spots. Acadia National Park is a mound of bedrock on the midcoast. The White Mountains form gray stepping-stones to Mount Katahdin, a mass of granite

139

at midstate. But the majority of Maine is pale green "till." That's a jumble of sand, silt, clay, and stones, which may, the map legend notes, contain many boulders. (Maine farmers would agree: Till is no fun to till, but it does make for picturesque stone walls.) And my yard, on the map, is pale gray, for "glaciomarine deposits."

My lawn's last dunking dates to fourteen thousand years ago. The advancing glacier had pressed the yard into the earth's molten basement, and when the ice retreated, the sea swamped the depressed land. The Atlantic overran Willard Beach, sloshed up the slope, and pushed halfway through the state. But you can't keep a good continent down. The crust rebounded so fast that by nine thousand years ago, the ocean had drained fifteen miles offshore from the current beach. It left behind mudflats spread like frosting over the lumpy till. This blue clay, studded with shells and worm fossils, turns up in backyards and gardens all over the southeastern third of the state. In my neighbor's new cellar-hole, it lies under three feet of sand that were probably laid down when the ocean was retreating.

After both ice and ocean cleared out of my yard, the southbound flow of life reversed. Tundra lichens and dwarf willows marched back to colonize the cold till. Blowing dust, plus dead plants and the corpses of their faithful insects, enriched the sand. Spruce trees caught this soil and thickened it with their needles and dead wood. Squirrels fed on the cones. Modern gray wolves twanged the tendons of caribou in my yard while the air grew warmer.

Over on the West Coast, the Bering land bridge was passable and populated with grazing animals. Two new species, man and dog, ate their way across the bridge, then poked their noses into the backyards of North America.

Is it a coincidence that a bevy of beasts promptly went extinct? The American mammoths and mastodons, camels and giant beavers, had evolved on a people-free continent. Their instincts may not have been honed against such a brainy predator. Some scientists contend that the yo-yo of climate change was more injurious, because it shifted plants up and down the latitudes too quickly for grazing animals to adapt. Whether North America's large mammals

140

starved or fed the newcomers, they died out at about the same time that dogs and their people swept through the Americas.

❧

The wolves in my yard must have raised their muzzles and sniffed hard the first time they saw dogs, not to mention the bizarre creatures that walked beside the dogs—walked upright, like plucked crows. Both new animals probably reeked of stale smoke and dead wildlife. Actually, gray wolves and dogs may have moved toward my yard at the same time, once retreating ice and sea made room for a food chain. They may even have overlapped with the fading dire wolf, if he, too, drifted north with the returning forests.

When this happened is controversial. Who crossed the broad expanse of tundra from Asia to Alaska, and when? Did they have to wait for the glaciers to part, or did they skirt the coast? On foot, or in boats? Did some of them hop across tropical islands farther south? Evidence of the stone-and-bone variety argues for a few pulses of immigration across the northern route. But evidence is sparse. The sea has risen again since then, swallowing the coastal sites where people would have left their garbage and broken tools. Bits of evidence hint that people may have peopled the Americas as early as thirty thousand years ago.

In my yard, the glacier simplifies history. People couldn't have been here until there was land to inhabit, about thirteen thousand years ago. The first sign of them turns up a thousand years thereafter. But they, too, are a hard bunch to read, says David Sanger, an anthropologist at the University of Maine. They left behind Clovis points—stone spearheads—and not much else.

"We have fluted points from a hundred different sites," he says. "But we have only a half a dozen good artifacts with those points. Did they have a caribou diet? Well, I'm sure they ate whatever they could find," he says with a helpless shrug. As for proof, bones are prone to dissolve in Maine's acidic soil. And pottery, which is more durable, hadn't been invented here yet.

141

Sanger has come down to the coast from Orono to peruse my neighborhood from a Paleo-Indian perspective. He's a tall, soft-spoken man, with hair, glasses, and clothing all in shades of silver and beige. A trace of an English accent peeks out of various words. He says that one thing the Clovis points do confirm is that the first Indians in Maine knew their rocks. Almost without exception, the points are made from chert, a stone that takes a sharp edge. It's a stone that's uncommon in Maine. So the folks in my yard also knew their trade routes.

"If you're living here," Sanger says, tapping an aerial photograph of my neighborhood, "how are you going to get your hands on that kind of rock? You might make periodic trips up to Munsungan Lake, where there's an outcrop. But that's three-hundred-something miles. Now, say your in-laws live in northern Maine, and you have periodic reunions with their band. They might bring you something neat that you can use, as a gift—like my mother-in-law might bring me a pie." Stone from five different states turned up in one Paleo-Indian site. These people had a lot of family reunions. But the family that chipped chert together vanished together. The Paleo-Indian culture blinks out after just fifteen hundred years.

The Archaic period from ten thousand to three thousand years ago is even more obscure. At first, the tools are made from crummy local rock, ground and pecked into shape. "It can be made into tools, but it isn't pretty," Sanger says. "They probably used points that were organic—antler and bone." The people were still mobile. But instead of trading rocks, they were trading . . . squashes? A site on the Piscataquis River has coughed up a six-thousand-year-old fragment of gourd. That family of plants was domesticated in Central and South America a few thousand years earlier and flourished in southeastern North America perhaps five thousand years ago, so this fragment has quite a story to tell. The new folks also left behind remnants of a fish trap in the mud of a lake.

I'm frustrated by the fact that the archaeological record is skewed toward people living in the interior of Maine. Here on the coast, people may have been writing $E = mc^2$ and eating soufflés six thousand years ago, but I'll never know about it. At that time, the glaciers were still melting, so the sea level was still rising. If a coastal Archaic person forgot his novel at the beach six thousand years ago, today that book lies a good five miles offshore. The rising tide slowed five thousand years ago, preserving only the most recent coastal sites for archaeologists. Those sites tell of a culture with sufficient maritime savvy to catch a swordfish, and the culinary sophistication necessary to appreciate foot-long oysters, whose shells are preserved in middens beside the Damariscotta River fifty miles up the coast. Although the interior and coastal people pursued different dietary paths and used different tools, Sanger says they were united in one spiritual practice. For most of this period, people all over New England buried their dead in distinct cemeteries, gave them ritual tools, and powdered them with red ocher. Around here, it was the first evidence of what Sanger refers to as "a cosmological and spiritual centrality."

But contrasts between the interior and coastal cultures persisted. Consistently, coastal people twisted their string in one direction, while inlanders twisted theirs in the other. And based on limited archaeological evidence, the coastal people—gasp!—allowed their dogs to chew animal bones. Among many North American hunters, that was taboo. The explanation was that it insulted prey animals, who would henceforth hide themselves from hunters. But in coastal sites, bones of deer, beaver, and other animals appear to be dog-gnawed.

Sanger turns his attention on an aerial photo spread on my kitchen table. He taps the dark ponds I found in the woods a couple blocks inland from my house. "These are where you'd find edible cattails and lily bulbs. It would also attract ducks, snakes, muskrats, turtles, and deer that would come to water." Then he fingers the beach. "At low tide you'd have clam flats." He grins. "Clams don't get no respect. There's not much to them besides water. But women, children, and even old people can gather shellfish. I like to imagine people grumbling about 'clams again.' But they're dipping them into fatty deer broth." He studies the map again. "You're protected from northwest winds. You can beach your canoe and be pretty close to the water." As for lobster, currently the most coveted food in Casco Bay? Sanger has never found lobster shell in Indian garbage heaps. Yet I've read European accounts of Maine Indians harvesting

143

lobsters that weighed twenty pounds. They dried them like jerky and also used them for fishing bait. (The European settlers fished with lobster meat, too, and used them for garden fertilizer when the crustaceans washed up after a storm. Today, we use fish for lobster bait. Times change.)

The most recent Indians in my neck of the woods were, according to the reports of European explorers, the Almouchiquois or Armouchiquois. In contrast to Mainers to their north and east, these folks were farmers, which afforded them a sedentary, town-building lifestyle. This didn't endear them to their neighbors, with whom they feuded. *Armouchiquois,* according to the best guess of historians, is a derogatory term that means "dog people."

~

Enough about people. What were their dogs like?

They may have looked like Carolina dogs. The Carolina dog was discovered only recently, running wild at the isolated Savannah River plutonium enrichment site. For decades, people in Georgia, South Carolina, and Florida have rescued the toast-colored puppies from roadsides, without giving much thought to their origin. Now the dogs are being bred intentionally. When I set eyes on my first Carolina-dog photograph, I'm a goner. The United Kennel Club specifies that a Carolina dog should bear an expression of "softness and intelligence." And they do. They're lithe creatures with huge eyes. The muzzle is delicate. The big ears stand up. The legs are long, built for sprinting after squirrels and pouncing on frogs. The tail ends in a quirky fishhook. Overall, the dog looks like a dainty dingo: same white toes and tail tip, same tall ears. DNA analysis, superficial so far, encourages the fantasy. The Carolina dog may be an archaic *Canis familiaris* who, like the dingo, the malamute, and a few others, is minimally diluted with European dog genes. Like other primitive dogs, Carolina dogs dig dens and regurgitate food for their pups.

When I search for historical accounts of Indian dogs, I find that European explorers often noted a fine muzzle, and an overall appearance that suggested a coyote or small wolf. Colors ran the gamut, as they still do in Carolina dogs. Then again, just as frequently as Euros described the Carolina-dog model, they described other things. In Marion Schwartz's *A History of Dogs in the Early Americas* I find reports of wolfy dogs, foxish dogs, coyote look-alikes, and the popular "short-nosed Indian dog," which was a foot-high model reported from coast to coast, and south to Peru. Often, I read of a tribe keeping two breeds of dog, one the size of a small terrier, the other comparable to a wolf or coyote. Numerous Europeans from numerous locations reported that Indians encouraged the larger dog to mate with wolves, which are stronger, pound for pound, than domestic dogs. Some observers commented that the big Indian dogs were distinct from wolves only in that they were silent or barked instead of howling. Among dog illustrations I find two 1840 etchings from a collection called *A Naturalist's Library*. The one titled "Dogs of the North American Indians" shows a pair of . . . um . . . Carolina dogs. Perhaps the head is a little broader, but other than that, even the soft and intelligent look is present. A companion etching with palm trees and cordillera in the background presents two more breeds. The yellow "carrier dog of the Indians" is large, and stout of leg, ear, and muzzle. The tricolor "alco" is one third as big, with folded ears.

If the European reports are accurate, the dog-human relationships of a few hundred years ago echoed the modern range. Many Indian tribes trained dogs to hunt—everything from elk to bison, turkeys to beavers, even seals, depending on the ecosystem. Some dogs were coddled for sacrifice to the gods. Many more were fattened for the family pot or the market. The Spaniards, Schwartz reports in his *History of Dogs*, ate oodles of these on their American explorations. At the other end of the spectrum, many Indians observed taboos against dog eating. They sheltered their dogs indoors and loved them like family members. Women sometimes nursed puppies at their own breast. Occasionally, archaeologists find dogs who were buried with as much care as were people. Sometimes a beloved dog was killed to join a master or mistress in the grave. In Arizona's White Dog Cave, mummies from about A.D. 100 preserve a family structure: Mummy man was found lying beside

145

a big, white dog; mummy woman was accompanied by a small, spotted dog.

But on the timeline of people-and-dogs in the Americas, most of this evidence is recent. The Asian invasion by people and dogs stretched over thousands of years, and any number of different lifestyles. The man-dog relationship from that period is a mystery. And shortly after Europeans discovered the Indian dogs, the Indian dogs were diluted nearly out of existence, leaving even less information to work with.

After Asian dogs and their people walked into North America, the dire wolf checked out. But the other canids stuck around, in what must have been a strained coexistence. Undoubtedly, Indians killed some wolves. And wolves ate some dogs, and dogs ate some foxes. Dogs may even have eaten a few wolves. Wolves probably ate a few people. In the West, coyotes joined the mêlée. It was a jittery community, but nobody went extinct—not until a new wave of people washed onto the continent from the opposite direction.

The easterly series of immigrations deposited European men and dogs in the Caribbean and the east coast of North America. This time, the dogs were trained to hunt men, not elk and rabbits. Spaniards sets their mastiffs and greyhounds on those Caribbean and American Indians whom they deemed troublesome, immoral, or simply expendable. Schwartz reports that Spanish explorers sometimes traveled with Indian slaves, whom they butchered for dog food. Depraved as the Spanish were, there was plenty of violence to go around. It was not a gentle time. Many Indian tribes were accustomed to bashing enemy women and children on the head. The removal of captives' ears for the captives' own consumption seems to have fallen within the rules of war for some tribes; and various European chroniclers reported cannibalism in some tribes. Many Indian tribes were as accomplished in the slaving business as the Europeans were. In my flaking, 1835 edition of Drake's *Indians of North America,* I read that European men did a brisk trade in young Indian women, while Indians enjoyed having a Frenchman around to fetch wood and water. But even in this harsh context,

146

those rumbling war dogs must have given Indians nightmares.

The Indian population soon crumbled to a shred of the pre-European number, through smallpox and other plagues. Many survivors fled the Great Dying that beset my neighborhood, moving east, or north to Canada. The remaining Indians and Europeans behaved peaceably for about fifty years. The wolves—and bears, too—refused, however, to check their murderous tendencies toward the Europeans' pigs. And the foxes were charmed, I'm sure, by their introduction to domestic fowl. The wild canids soon made the acquaintance of the trap and gun. Like the Indians, they began to retreat and die out. As for the Indian dogs in my area, they were probably accused of taking European livestock, as they were elsewhere. And they surely interbred with European farm dogs. For a while, though, the surviving Dog People and their dogs carried on with daily life, innocent of the fact that an English king had essentially deeded New England to a trading company.

But the Europeans spread. If the large animals of the ice age displayed a fatal innocence toward two-legged predators, the Indians showed a similar innocence with regard to real-estate predators. Up and down the coast now, Europeans were clearing and planting, logging and shipping, and allowing their cows to mash Indian cornfields. The Indians got the picture: These pink-cheeked critters with their four-legged livestock were going to eat them out of house and home. Hostilities broke out across New England. Indians torched farms. The French joined the Indians against the English. In my neighborhood, my Anglo predecessors scurried to a newly constructed fort. The French lured them out with a promise of safe conduct to a nice English town, but then hiked most of them to Canada and sold them, presumably as servants to French Canadians. Survivors retreated forty miles south to Wells. When the last homestead had burned, my neighborhood was its old self for a time. Pasture-pine and birches shot up in the fields. Ten years later, under one of many peace agreements, the English came back. Then five years later came more war, slaughter, and captive-taking.

So it went for sixty years, with alternating terror and peace. In my yard the politics likely played out as alternating pines and pasture, pines and pasture. But with each round, the Indians grew more hungry and less numerous. Their corn was vulnerable to European flames and animals. The Europeans refused to sell them powder and shot, for fear it would be returned to them at high velocity. Even dogs were withheld from the Indians. And they seem to have forgotten how to hunt without guns. The immigrants continued to gnaw into the woods, claiming, taming, and renaming.

It was a long struggle, with enough horror to scar a nation to the bone. In my neighborhood, as in most yards throughout the Americas, the Indians and their dogs lost. Once, they ruled my backyard. Now I can scarcely reconstruct who they were. The closest I get to their dogs is a painting from 1855. It depicts canoeing Micmacs, one of the Armouchiquois' feuding neighbors. Dozing outside the birchbark tepee in the background are two, well, Carolina dogs.

ᔥ

For many decades after the Indians were gone, my yard was probably a bleak pasture. In photographs from the mid-1800s, the land looks close-shaven, presumably by unfenced livestock. Trees are scarce. Houses, for that matter, are rare. They stand like clapboard boxes on a warped table. Dirt trails link them together. On a map from about the same era, my street doesn't exist. Just fifteen buildings are sketched on the six-block area where today more than a hundred stand. Most of the land, including my lot, is open. I suspect it was an orchard or pasture belonging to one of the Willard clan, whose various homesteads are nearest.

In Captain Benjamin Willard's autobiography, I find signs that my yard was already, in the mid-1800s, destined for suburbhood. In the illustrations intended to showcase the captain's trained pets (besides inducing Spot the dalmatian to climb onto his head, Willard trained a cedar-waxwing bird to perch on Spot's circus headgear), I find signs that the sun has set on the day of scratching a living from a wild land. One room is tricked out with elabo-

148

rate moldings. Wall-to-wall carpeting covers the floor. The captain himself appears sleek and well-stuffed, not a man of manual labor. His personal economy was so robust that he left a monument to Spot when the dog died. Spot went blind at fifteen, and a doctor etherized him. Then Captain Willard "had him buried on my lot . . . with a fifty dollar headstone." On an afternoon with low clouds and hardening ground, I search for this plot, up on Meetinghouse Hill. In the shadow of the captain's granite alp stands an arched slab. Carved on it is a dog, his forepaws engaged in rolling a barrel. "Trained Dog Spot," the stone reads. If the wolves who once howled in my yard could see what their tame cousins stooped to, they would surely curl a lip. But by Spot's time in the late 1800s, wolves had been hounded out of the state.

The wolves were part of a trend. The number of different animals and plants that could make a living in my yard had been shrinking since the first people and dogs sauntered in twelve thousands years ago. All the dire wolves, primitive horses, camels, and mastodons who chased the shrinking glacier north had gone extinct. Then, when the Europeans arrived, nearly every tree in the state was felled. In the open pastures, few plants could compete with the imported grasses that fed imported horses, sheep, goats, and cows. And few wild animals could survive on that carpet of grass. But when suburbs sprouted in the fields, this trend toward sterility stopped, then shifted into reverse.

Suburbanites wanted trees. They wanted flowers. And they either didn't want or weren't allowed livestock. Backyards became oases of diversity. Granted, as often as not the plants people embedded in their yards were native to Asia, not New England. But they looked good to many a hungry insect and seed-seeking mouse.

Bigger creatures adapted more slowly to the new landscape. Coyotes won't normally share territory with wolves. But people had pushed wolves deep into Canada and Mexico. The yellow canines came sniffing east. Nocturnal and audacious, they're concluding that life near people is tolerable. It doesn't hurt that suburbanites are forbidden to trap or shoot them. Now instead

149

of wolves howling over the settlers' pigs, coyotes yammer over the cats and dogs in the same neighborhood.

White-tailed deer take a similar view. As fewer people hunt, an exploding deer population is prancing deep into the suburbs, where food is abundant. Ditto crows. And gray squirrels. And the other night, a neighbor spotted a moose rambling through the graveyard where Trained Dog Spot lies. The resurgence of animalia puts me in mind of postglacial rebound. Farming and lumbering kept the land nearly empty of wildlife. Now drained of agri-industry, the land is filling with creatures again. The most recent native to investigate my area is the wild turkey. Hunted nearly to extinction, the portly birds now dally on suburban roads and raid bird feeders. A friend who works in a suburban office building e-mails me photos of them sitting on cars in the parking lot.

Even that old-time canine, the wolf, is drifting back toward town, striking out from a base near the Canadian border. In Maine, the first wolf to be spotted in a century was shot in 1993 by a bear hunter who may have lacked a clear concept of what a bear is. The next one got to the coast before it was trapped, then shot. It could be a slow return. But how interesting if they should survive the gauntlet of guns, traps, and coyote snares; if they should find enough to eat to support their trip south. Wolves lived cheek by jowl with Indians. Only relentless persecution by Europeans with cheap bullets, traps, and poison drove them away. Without that persecution, perhaps wolves, too, will find the suburban lifestyle to their liking. Stranger things have happened. In the suburbs of Boulder and other western cities, mountain lions recently discovered grass-fed deer, then diversified their diet to include cats and dogs, and even the occasional *Homo sapiens*. How interesting, to wake in the streetlighted suburbs to hear the silver wail of a wolf. To find those big pawprints on the first snow. To spot among the weeds, not a disemboweled sparrow, but the glossy black hoof of a deer, then the tawny leg.

Up the hill, Trained Dog Spot is surely rolling in his grave.

8

THE STATELY AND SCHEMING TREES

A GLINT OF penny-colored fur catches my eye. Lightning stripes give it away: It's Cheeky the Chipmunk, popping across the grass to his hole. Something protrudes from his mouth. My baby found an acorn, all on his own.

I've been worrying about Cheeky's diet. I read that sunflower seeds are the equivalent of junk food for chipmunks. But I can't convince him that other foods have much merit. I thought that when the weather cooled, his little body might demand more serious nutriment. The only thing that changed was his work schedule. Instead of *pittering* into the house at eight A.M., he turns up at nine, then nine thirty. On an overcast day, his claws might not hit my knee until eleven. But there he is now, working independently, and diversifying his diet. I look out my office window later in the day to discover him at eye level, twenty feet up in the chokecherry tree.

I miss the opportunity to smooth his lightning stripes and smell the herbs in his fur. And I hate to see him bopping around in the open, exposed to cats and hawks. His path to the indoor seed dish is well shielded by shrubbery. Still, I'm glad he's able to break a habit when instinct hollers at him. I'm also proud that he's able to find acorns in a year when the oak trees are making a coordinated effort to starve him out.

They really are. It's not accidental that the acorn crop is piddling. The trees have put their crowns together and decided to thin the rodent population. Throughout the neighborhood, and perhaps throughout New England, oak trees are holding back. Each tree is producing a minimal number of acorns. Some are producing none at all. As a result, the deer, bears, squirrels, and birds that depend on acorns are planning famine-size little families, or none at all. If every single acorn is devoured by hungry critters this year, the trees can tolerate that. They have decades to reproduce. And one of these years when the acorn-eaters are trimmed down to size, the trees will wink, nod, and dump down a monster crop. The diminished animal population will be unable to harvest them all.

When I heard about this "masting" strategy, I found it suspect. Are we to believe trees have a secret communication system, or an OPEC-style acorn cartel? But judging from nascent research, it's simpler than that. The trees have evolved to respond to weather. A certain pattern of drought and heat presses the right chemical buttons in the trees, and they go nuts with the acorns. Evolution rewarded the trees that chanced to participate in the cycle. Trees around the world have evolved to respond to various weather patterns. Trees in eastern North America, for instance, synchronize their fruiting clocks to an early-summer drought. The next year, they let fly. Cooperation between different species also varies from one area to another. In an extreme case of synchrony, almost four-hundred different species of trees in Borneo rely on a cue from El Niño, whereupon they all bombard the seed-eating boar, orang-utans, parakeets, and people with a storm of seed.

Compliance with the starve-'em strategy isn't perfect. My friend Deb, who lives a half mile west of me, has been made sleepless this year by the clang of acorns on her car. Her driveway is paved in yellow pulp. Her tree is masting this year, and damn the conse-quences. Perhaps it's a mutant, or perhaps its genes that respond to its personal water or food situation overpowered its genes that act for the greater good. This happens in a wild forest, too: Not every tree plays along. But enough of them do so that synchro-nized masting works.

In the interest of full disclosure, this "predator satiation" theory isn't the only one on the menu. One competing hypothesis argues that the trees produce high mast after they've been perfectly fed and watered. Another says that trees who depend on wind for pollination need their comrades to be in bloom at the same time so that they can cross-pollinate, and that pollination drove evolution toward synchrony. I wouldn't be surprised if all three factors play a part. But I also wouldn't blame the oaks if their primary objective was to defeat the acorn-stealers.

In a low-mast year like this my red oak might produce a few hundred acorns. Each has the potential to become a new oak tree. None is likely to succeed. An acorn is an attractive package of calories, low in protein but high in fat. An ark's worth of animals await their ripening each year. The 150 species of acorn burglars include acorn weevils, mice, blue jays, woodpeckers, turkeys, foxes, feral pigs, deer, and bears. In the suburbs and cities, birds and small rodents are the heartiest eaters. Of all these, only a few pay for their meal. Squirrels and jays bury an acorn in a way that maximizes the nut's chance of germinating. These critters intend to dig up the nuts later. But if they forget, or die, the oak benefits. By contrast the weevils, the deer, the woodpeckers, and the bears either chew up the nuts on the spot or stow them in tree cavities or rain gutters where they can't grow. The truth is that suburbanites are happy to have squirrels and weevils constrain oak fecundity. When we desire another oak tree, we'll go buy one and put it where we want it. But my tree's genes urge it to reproduce, so it soldiers on.

The masting strategy generates unexpected ripples through an ecosystem, beyond regulating the squirrel population. A recent investigation found that a high-mast year results in a higher population of deer and mice; in turn, that produces a bloom of the ticks that carry Lyme disease. On the other hand, a low-mast year trims the population of mice, who eat gypsy moth pupae. So an acorn shortage can lead to a gypsy moth surplus. Nature is complicated.

The other trees in the yard are having as lean a year as the oaks. The few apples that the orioles spared would benefit if someone ate them and then excreted their seeds wrapped in a blob of fertilizer. The fruits advertise their sweetening flesh by replacing green camouflage with reds and yellows. But there are no takers. The apples rot where they fall. The sumac trees are barren, too. Thousands of rose-chafer beetles held an orgy in the blossoms earlier in the summer, devouring every flower. Not one berry materializes. Oh, well. A tree has patience.

They're embattled organisms, my trees. Their fruits are stolen, and day after day, their leaves are eaten alive. They're absolutely crawling with herbivores who want a piece of them. I climb a stepladder into my oak and perch among the low branches. At first glance, the leaves are a rich forest-green, laced with bright veins that put me in mind of lighted highways seen from an airplane at night. The same principle creates both patterns, I suppose: A main artery divides and divides again, to deliver resources to the humblest town or leaf cell.

Subtle motions draw my attention. On the leaf surfaces, white animalcules the size of salt grains move about. Are they eating leaf, or eating mold that's eating leaf exudate? A glittering little fly I think of as a greenie is gliding across leaf after leaf, in search of aphids. A small orange spider hangs in a web she has stretched across a rosette of leaves. I swat a mosquito and try to make a gift to the spider, but either my looming hand or the mosquito corpse terrifies her and she plunges down on an escape thread. The mosquito revives and kicks free. The spider returns. Out at the end of this branch is a Zippytown ant, a midsize brown gal. She's hunting, exploring every twig she encounters. She has commuted twelve feet up the trunk, and another ten to the end of the branch, in her quest for food. If I can spot this many animals on a single branch of the oak, imagine how many others are scuttling around me. This tree is a world unto itself, a skyscraper bustling with insect industry. A catbird alights six feet away, chattering full tilt. His gray chest heaves. He catches sight of me through the leaves and claps his bill shut. He peers with one black eye, then tilts his head and peers with the other. Then flies off.

The oak leaves, on close inspection, are battle-weary. I count the bite marks and holes on a few of them. The average number of insults per leaf is 350. Every leaf in my sight is battered. Some

154

are gnawed right to the spine. Every leaf stem has been chewed, too, and scarred with black measles. The tree is being eaten alive.

Unlike animals, trees can't get up and run when an enemy approaches. And my fifty-foot oak can't hide. It's a rooted tower of food. Back in May when its leaves were pale green stars, they were irresistible. In the woods, deer and rabbits would have made short work of any low foliage, but those animals are absent from my yard. Instead, the shiny leaves were nibbled by squirrels. As the leaves toughened, the tree was invaded by a horde of insects. Gall-making wasps and mites deposited eggs on leaf and twig, inspiring the tree to grow an edible uterus of oak cells. On the leaves a splendid array of beetles, flies, moths, and wasps left eggs that produced young leaf miners. These larvae spent their days tunneling between the upper and lower surfaces of the oak leaves, their paths drying to brown squiggles. Even under the bark, a beige oak-borer larva awoke and resumed his conversion of hardwood to sawdust. Tiny gypsy-moth larvae chewed holes in the new leaves; as they grew and molted, they ate bigger holes.

My oak tree didn't take the herbivore attack lying down. To experience its first line of defense, I take a few bites of oak leaf myself. It's hard to find one that still has a bite-size section free of holes, bumps, and brown scars. And for good measure I wash the leaf. Then I snip out a nice section and chew. Initially the flavor is powerfully green, and not odious. But as the leaf turns to mush, it turns bitter. It's not puckery, nothing like dandelion. There's a nice note of black tea. What's bothersome, though, the reason I wouldn't serve steamed oak leaves as a side dish, is that it does something to my saliva. My spit thickens, I think. It feels super-slippery. And the inside of my mouth feels a little weird. Hmmm. Maybe it's time to try an acorn.

I find a nice one, with no tooth marks or weevil holes. I pry off the cap and wash the nut. I don't want to introduce the confusing symptoms of botulism to my experiment. Oops. Didn't notice the weevil hole. That one's black and punky inside. As is the next one, and the next. Well, I've been foiled by a low-mast year. Weevils can get almost every acorn in a slim year. They must have missed a few, because the squirrels got some, and they won't take breached nuts. But between rodents and insects, no acorns are left for me. I do remember biting into them in my childhood, though, and the memory is of a desiccating bitterness.

Both the leaves and fruit of an oak tree are drenched in tannin compounds. Tannins taste nasty, but that's not their best feature. As I chewed up the plant cells, little packets of tannin broke and mixed with the pulp. Tannin molecules sought out nutritious protein molecules in the leaf pulp and bonded with them. My body would have a hard time absorbing those altered mol-ecules. I suppose that reaction also altered the texture of my saliva. And perhaps that alteration cued me to spit the stuff out. I had hoped to swallow the mush to see what happened. But the reptilian portion of my brain may have interpreted my spit-thickening as a warning of indigestibility. It simply refused to let me swallow. I doubt the oak leaf would have hurt me. Some insects do grow slower when they feed on high-tannin vegetation, however. While that's not a fatal blow, it's a stealthy strategy on the part of the tree: Slowed growth forces larvae to spend longer feeding, which exposes them to birds and other predators. Heh, heh!

Now, if tannins deter some feeding, why don't oak trees produce more of them and scare off every diner? For one thing, making tannins burns a tree's energy, energy that could be channeled into acorns. For another, a tannic acorn may offend an acorn weevil, but it will also offend a squirrel. Squirrels will accept high-tannin acorns in lean times, but they prefer mild ones. If an oak tree gets too heavy-handed with the tannins, the squirrels who plant its acorns may patronize a sweeter tree. The oak must use its defenses judiciously to maintain congenial relations with its seed-planting partners.

Plants who need no help with their seeds are free to be meaner. The milkweed that grows near my raspberries, for instance, uses wind to distribute its seeds. Accordingly, it can deploy a nastier chemical, designed to jangle the nervous system of its attackers. The white sap in milkweed not only tastes bitter, it also contains alkaloid chemicals that can give an insect a heart attack. Even animals as big as cows occasionally overdose on milkweed. The dose makes the poison, however, and human beings have learned to love many plant alkaloids in measured amounts. Tobacco plants make alkaloids. Coffee trees do, too. Poppies and coca plants gush alkaloids. We intentionally extract those chemicals to jangle our nervous systems. Nicotine, caffeine, morphine, and cocaine are among the most entrancing chemicals we know.

The pine trees in my yard subscribe to another common class of chemical weapons, the terpenes. Some terpenes kill a feeding insect outright. Others scramble a larva's development by sterilizing it, destroying its appetite, or making it molt prematurely. As with alkaloids, people have found ways to harness terpenes. Pines give us turpentine, pine oil, and rosin. We use camphor for the same thing the camphor tree does: to repel insects. From mint we get the terpene menthol. The soapwort I used to wash my hands when I was investigating my lawn herbs uses its soapy terpenes to poison a host of beetles, mites, and moth and butterfly larvae. And so on. Most plants have more than one chemical defense. Tannins turn up in many plants, as do the soapy chemicals.

Some of my plants also rely on mechanical defenses. The oak, in addition to its tannic bite, offers wood and leaves that are toughened by nutrition-free fiber and lignin. If an animal spends more calories chewing its meal than it gains from the food, it'll soon find something more profitable to eat. The pines use pitch to catch the insects who puncture their bark. The raspberries fend off herbivores with thorns. *Nicotiana,* the tobacco plant, deploys its toxic nicotine from spiky hairs that defend its leaves.

My oak, in addition to its chemical and mechanical defenses, may also be part of a community-action group that fights germs together. A plant that's bitten or infected releases a burst of special chemicals into the air. If my oak can detect a fellow oak's distress signals, then it will take heed and harden its own defenses. Should the plague of fungi or caterpillars attack anyway, my oak my issue a different chemical bulletin, a "soup's on" alert to lure the attacker's own predators. That's pretty clever, for a big stick.

Most experiments on "talking trees" have been conducted on small, manageable plants like tobacco and sage. But researchers increasingly suspect that many plants can smell when their neighbors are battling bugs and react to protect themselves. In one of the few experiments with trees in the open air, researchers played the role of herbivore in a stand of European black alders. They ripped leaves off some trees, which theoretically stimulated the alders to let out a chemical yell. Researchers then monitored the amount of natural attack borne by neighboring alders, whose leaves were not ripped. The trees nearest the wounded alders experienced the least gnawing by bugs. Presumably, they were able to detect the chemical yell and pumped their own leaves full of defense chemicals. Experiments with smaller plants have demonstrated all of these steps: A torn sage plant releases a burst of methyl jasmonate. A tomato or a tobacco plant downwind from the injured sage plant boosts its own defensive chemicals. And those downwind plants save themselves from herbivore attack.

Plants can be surprisingly sophisticated with their shouting. In one field trial with tobacco plants, researchers set tobacco budworms on some plants, and corn earworms on others. Each herbivore solicited a different mix of twelve chemicals from the plant it bit. Right on cue, a species of parasitic wasp that preys on budworms showed up, and those wasps flew to the budworm-infested plants. In experimenting with the yell chemicals, scientists have noticed that wild tobacco also seems to release a nighttime burst of repellent as a matter of normal hygiene. It appears to prevent moths from laying eggs on them. The combined effect of a plant's chemistry set is impressive. In one study, scientists calculated that a plant rebuffed over 90 percent of possible insect attacks.

Plants may even communicate about bacterial infections. Many, when their leaves are injured, flood their tissues with a blast of oxygen to burn invading microorganisms. But they also release a cloud of antibiotic vapor into the air. Scientists haven't determined if nearby plants take the vapor as a warning. It's a new field of research, and the uncertainties are large—which is why I can't be sure whether my oak tree is a communicative genius or a stupid log. It seems likely, though, that the air around me is rife with chemical telegrams. It doesn't seem far-fetched that one insect's saliva chemical could induce one chemical reaction from my oak tree, and a different saliva chemical could elicit a different reaction. My own immune system does something similar. Nor does it seem amazing that a predatory wasp could evolve to perk up when she smells a certain distress chemical emanating from oak leaves.

So my trees are not helpless. They're armed to the teeth, and they have a lot more time to reproduce than their insect enemies have. If they're healthy, they can even weather the loss of an entire set of leaves to gypsy-moth caterpillars. And sometimes it comes to that. In the evolution arms race, my oak tree has won some battles, but the war never ends. Mutations happen. Sometimes the mutation hands the tree a new defense; other times, it bestows a new tool on an herbivore. The herbivorous gypsy moth is currently one mutation ahead of the oak tree.

Some years ago when I lived in an oaky neighborhood, the gypsy-moth caterpillars had a boom year. They wiggled in platoons across the driveway and up the trunk of the tree by my house. And they ate. The trees they'd ravaged on the other side of the driveway held up leaf skeletons against the blue sky. The leaves on my tree were soon converted to caterpillar meat, and to frass, which pelted down in brown nuggets to cover the house, the car, and the lawn. The snicking of caterpillar jaws was audible. When my tree was eaten back to gray twigs, the caterpillars moved on.

What gives? Shouldn't a gypsy moth starve when it spends all its energy eating inedible tannins? Scientists don't yet know exactly what method the caterpillars use to prevent tannins from binding up every shred of nutrition in an oak leaf. Whatever it is, it's not yet perfect. Gypsy moths that feed on oak trees grow more slowly than their kinfolk who eat other plants. But Jack Schultz, from a research team at Penn State, told me he thinks the caterpillars have found two ways to defang tannins. One, the larvae keep their guts unusually well-oxidized, which makes the tannins stick to each other, rather than to proteins. Two, the tannins also get stuck to fat molecules in the larvae's guts. While this deprives the caterpillars of some fat, it does permit other nutrients to be absorbed.

Other acorn eaters have out-mutated the oak, too. Squirrels have evolved to eat only the least-tannic part of an acorn, if nuts are plentiful. They'll bite off the top and leave the rest. Deer have evolved special spit that combats tannins. Humans have, too—my oak-leaf aversion notwithstanding. In my saliva, special proteins bind up the tannins and escort them through my body, which prevent them from hijacking the acorn protein. It's only logical that we would have evolved coping skills, since tannins are so common in the plant world, and since plants are so common in the human diet. To manage the supertannic acorn, though, humanity has relied on technology. Many American Indian populations pounded acorns into flour, then soaked out the tannin.

Sometimes an animal who defeats a plant's chemical defense evolves further to exploit those chemicals for its own use. The monarch butterfly is one. The monarch caterpillar eats the heart poisons from the milkweed plant, then sequesters them in his body, which makes him taste terrible to birds. (Like many creatures, the monarch publicizes a foul flavor with bright colors, as if to say, "If I'm this easy to see, there's probably a catch." The viceroy butterfly, an evolutionary cheater, mimics the monarch's colors and forgoes the bitter diet.) The tobacco hornworm does likewise with the toxic atropine in tobacco. Instead of succumbing to atropine poisoning, the green giant stores it inside to repel her own enemies. And the gypsy-moth researchers at Penn State say that gypsy-moth caterpillars can self-medicate with oak tannins to protect themselves from the deadly "wilt disease" virus. They dose themselves with the same chemical intended to send them packing. Even we humans are appropriating the oak's intellectual property and testing tannins for antiviral properties.

This year my oak escaped major plagues. I'm not sure why, because the oaks in a park a few blocks away were ravaged by a horde of oak skeletonizers. Tiny moth larvae are currently descending from those treetops on threads of silk, heading for a winter in the duff. They dangle in such profusion that the edge of the oak stand glistens like a curtain. From my tree, only a couple of streamers wave. The leaves are dark, still hauling in sunlight. I can't prove it, but my personal theory is that my tree has a good nose for trouble.

Backyard trees face special challenges related to their urban lifestyle. But they reap benefits, too. Paradoxically, town trees are often the tallest around. Just ask my crows. Kevin McGowan, the Cornell crow man, has proposed that the size of the trees is one of the amenities that tempts crows into town. Initially, I thought it odd that town trees would be bigger than free-range trees, but there are logical reasons for the difference. One is that town trees don't get chopped down to make paper and plywood. These days, some of the oldest old-growth forest grows in parks and old neighborhoods. Street trees and backyard trees also enjoy reduced competition. They're seldom crowded together to fight over light and water, as forest trees are. Like my oak, a city tree can grow a broad, efficient crown. Being isolated from its peers, it's also less likely to be caught in a plague of caterpillars, or a viral epidemic.

The town tree even gets a helping hand from the town's dirty air. My car engine puffs nitrogen (and a zillion other things) into the air as I go about my errands. Nitrogen can be a poison or a fertilizer, depending on your perspective. From the poison perspective, the nitrogen joins droplets of atmospheric moisture, which becomes acid rain. That harms plants. And nitrogenous rain can feed the algae in lakes and ponds, which stresses other water dwellers. But nitrogen is also the classic fertilizer chemical. To a tree like my oak, who gets no fertilizer from me, nitrogen that arrives in the rain may be just what the doctor ordered. My tree can even get this shot in the arm from a rainless wind, by straining out droplets too small to fall on their own. When flowing air slides up against that wall of oak leaves, the gas molecules change direction and climb over the tree. But microscopic droplets splatter.

161

Scientists from the Institute for Ecosystem Studies in Millbrook, New York, have measured the amount of airborne nitrogen that splats on forest edges, versus the interiors. The bonus at the edges is huge. Those trees get one and a half times as much nitrate as the interior trees. The research was done on big chunks of field and forest and focused on the "edge effects" produced by a line of trees, rather than a singleton. But smaller edge effects are rampant in the cities and suburbs, where microforest meets microfield. Every edge forces the wind to switch direction. Of course, this nitrogen fertilizer from tailpipes is only a boon if the lawn isn't already stewing in commercial plant food. The dose makes the poison, which is clear to anyone whose lawn has dead spots from nitrogen-rich dog pee. Too much of a good thing is a bad thing.

Ozone pollution also works in a town tree's favor, if in a back-handed way. When I came across a study in *Nature* claiming that ozone causes young city trees to grow faster than country trees, I was doubtful. Ground-level ozone, formed when my car exhaust reacts with sunlight, is good for trees? I had thought it burned their leaves the way it burns human lung tissue. On a closer reading, I found that ozone is indeed bad for trees. But the reaction that turns exhaust to ozone takes some time. So when I start my car, my exhaust blows harmlessly past my oak tree. Sunlight works on the chemicals as they drift on the wind, out of town and up the coast. By the time my exhaust filters through a stand of seaside spruce, it's ripe and reactive. The wild trees bear the brunt of my city driving. They grow more slowly. It's not that city air grows healthy trees. It just grows less-sick trees.

A tree's life on the lawn does have drawbacks. Urban air carries soot and other dusts that jam open the stomate pores of a leaf, which prevents them from closing to hold in moisture. City air is also a few degrees warmer than rural air, which causes trees to release more water to cool off. On the bright side, the heat can stretch a tree's growing season.

Urban soils also handicap trees. The soil through which their roots chase nutrition is often junk. The flavor of the garbage varies from place to place, but common varieties include acid soil from acid rain; alkaline soil from the disintegration of concrete side-walks and foundations; compacted soil from feet and tires; water-proofed soil from the wax trees secrete on their leaves in response to pollution; and microbe-poor soil that produces tree food slowly. One of the saddest deprivations town trees weather is the loss of some little buddies who ought to be growing along their roots. Mycorrhizal fungi are symbionts who web out from tree roots to do the yeoman's labor of hunting down nourishment. In addition to feeding their tree, they also seem to protect trees from drought, and they condition the soil into the perfect size of crumb. In exchange for all these services, the tree sends carbohydrates down to feed its fungi. Various investigations have found that town trees lead a rather lonesome existence in this regard. They may be accom-panied by some fungi, but not the diversity of species that rural trees enjoy. Lots of other plants also get assistance from mycor-rhizal fungi. But trees need the fungi so badly that you can buy them in a can, to augment urban dirt. The durability of this cure is questionable. Science suggests that the fungi may need clean soil if they're to stick around.

Despite its mistreatment, my oak tree offers goods and services worth thousands of dollars. Not least are its aesthetic labors. By mid-October, trees across the state are closing down for the winter, and leaf peepers are spending hundreds of millions to watch this colorful contraction. My own yard is like a painting that evolves every day. A few forsythia leaves shift from a lime color to deep purple. The sumac leaves grade from green to yellow, then orange, then scarlet, blazing like red stars. On the chokecherry, the leaf hues are an assortment of peach and avocado. In the Bamboo Wilderness bittersweet vines turn lemony. The oak is a late bloomer, offering only a few red leaves. It will hold its dead, brown leaves until a November storm flails them away.

The foliage display of northern trees was long thought to be a side effect of the trees shutting down their energy production. As the green chlorophyll degrades in the leaves, yellow and red pigments that were there all along come to light. So went the theory. But I've recently come across a few papers suggesting the color has a function of its own. One of the clues is that red "antho-cyanins" don't flood leaves until late summer. And anthocyanins function as sunscreen, antifreeze, and antioxidant. One theory, then, is that the red chemicals prolong the life of an old leaf, giving it more time to drain its nutrients into the tree. A second theory is that trees use color to warn away enemy insects seeking a tree to host their eggs for the winter. The healthiest trees make the brightest leaves, this theory goes, signaling to insects that their kids will meet a vigorous chemical defense when they hatch and start feeding in the spring. One paper has found a correlation between the brightest fall trees and the least insect damage the following year. But a correlation isn't proof of cause and effect. The red-leaf mystery isn't resolved. Whatever their reason, the deciduous trees in my yard spend a few weeks making themselves gorgeous, before they shed their leaves altogether.

My oak tree performs practical tasks, as well. In summer, its leaves prevent the sun from beating on my house. That reduces my need for air-conditioning. Its crown casts a shadow within which it can be thirty degrees cooler than the unshaded lawn. My tree also soaks up dangerous pollutants like ozone and volatile chemicals and neutralizes them. At this very moment the oak is trapping microscopic dust particles that inflame human lungs. And its growth is soaking up carbon dioxide from the atmosphere, countering what my car emits. Although the carbon stored in the leaves will rise to overheat the globe again when the leaves decompose, the trunk and branches are taking carbon out of circulation for centuries. The tree is even transferring some carbon down into the soil for long-term storage.

Lately some tree-lovers have attempted to put a dollar figure on the work city trees do. Accounting methods vary, with some studies looking only at the value of reducing pollution, and others trying to tack a figure on the aesthetic value of a tree. A Chicago tree, for instance, is judged to save $402 a year in pollution-abatement and air-conditioning costs each year. For every dollar spent on a tree in Modesto, California, the tree will return $1.89 in air-cleaning and cooling services, and by raising property values. In

164

New York, a small honey-locust tree earns $637 a year through pollution-control and landscape value. An enormous tulip tree earns $23,069. A big tree like my oak, with its thousands of leaves, is sixty to seventy times better at air-cleaning than a small tree. On that basis, my half-grown oak might be doing $10,000 worth of pollution control and beautification this year.

But that's still a narrow view of a tree's worth. My pines, for instance, stand between the house and the north wind, reducing my need to burn fossil fuels in winter. The oak obligingly drops its leaves just when the house is in need of solar beams in winter. My trees release oxygen, which I inhale. When it rains, the trees slow the drops, increasing the percentage of water that sinks into the soil. Their leaves, both on the branch and on the ground, muffle the sounds of humanity around me. The terpenes that my pines release on a hot day give the air a priceless spice. The real value of my trees is too complex for me to characterize. I wouldn't want to live without them.

None of these "value your tree" studies considers the value of a tree to animals, either. To my birds, trees are insect cafés, resting spots, and sentry posts. The crows even derive recreational value from the apple tree, which offers a bumper crop of dead branches to yank, peck, and twirl in the beak. The squirrels rely on my trees for protection, scrabbling up the nearest trunk when they're alarmed. They also make their leafy dreys and their baby-nursing nests high in the trees. My young pines are too short to offer much besides the usual complement of insects, but the neighbor's pine has sheltered a nest of crow eggs in the past. The bleached sticks and grasses of an old nest are spilling slowly down the trunk. Fruit from my oak and apple trees feed everyone from mice to microbes, plowing nutrients back into the yard. And, of course, the trees make chipmunk food. Not that I have need of that by midautumn. Cheeky is here one day, and the next day he is not.

He's not hibernating. Chipmunks don't hibernate, and anyway it's too warm for him to begin a winter cycle of long naps. Cheeky has not relocated. Chipmunks don't walk away from their investment in a tunnel system and a stuffed larder. I'm also troubled to note that I have seen no sharp-shinned hawk haunting the neighborhood of late. On the list of possible explanations for my chipmunk's disappearance this leaves cats. Once again I seem to have lost a friend to a neighbor's pet.

For three weeks I continue to leave the back door open, and a cup of seeds on Cheeky's stool in the kitchen. Every time I pass the cup I check to see if anything's missing. When the weather turns raw, I close the door, but leave the seeds just on the other side. I open it frequently, to check for missing seeds. Whenever I cross the yard, I stop by his main hole and tell myself that the apple leaf at the entrance has a fresh pawprint on it, or that it has moved a half inch. The spicy scent of yarrow in the lawn, or of sunflower seeds, evokes a memory of the way his fur smelled. I miss the heft of him in my palm, and the warmth of that coppery velvet against my cheek. But Cheeky never comes again.

He was the only chipmunk I've ever seen at this house. Given the number of other people's cats who trespass in my yard, maybe that's not surprising. My yard may also be deficient in the ideal chippy food. Researching the relationship between trees and rodents, I discovered that while chipmunks certainly welcome acorns, they are even more delighted with red-maple seeds.

Come spring, I'll be planting a maple tree. It will muffle noise and mop up pollution, it'll reduce water runoff, and it might even look nice. Who cares. I just want it to feed the chipmunks.

9

SIFTING SECRETS FROM THE AIR

I THINK THIS is the first autumn that I've ever been happy to see tree branches turn bare. For months it's been a struggle to watch what's going on in the yard. Insects molted and matured, birds nested and fledged, squirrels built summer beds, all behind a screen of leaves. Now I can see again. On the morning after the oak relinquishes its garnet leaves, I find the crows lined up on their begging branch. How many mornings have they sat there, hidden by leaves, wondering, where's the dimwit with the dog food? Happy to see them so sharp and glossy against the sky, I rustle up their grub. When I step out the door, cold air makes me cough. The sky is so clear it looks as though a crow could shatter it with a peck. Before I set foot on the lawn, I sense that frost has stiffened the soil. Maybe sounds are bouncing differently across this rigid world. Even my startled nose detects a change. Gone is the mélange of industrious greenery and toiling soil. In its place is a delicate odor of dryness. The cold night to come is even more brittle. The stars blaze clear and white. A blowing leaf rattles like ice on a windowpane.

Winter is in the air. What does that mean? I wonder. What does air do so differently from one season to the next?

❧

Not so long ago, the air carried enough heat and moisture to drive itself mad. In July, I cowered on the deck as air boiled into thunderheads. A wedge of cold air was plowing south from Canada. It rammed under the local warm air, boosting it skyward. The displaced air churned and roiled, until it built a load of ice balls and electricity it could no longer contain . . .

At two on a sweltering afternoon, the world goes dusky. From the northwest, thunder grumbles. Hot wind rushes around the yard like a rabbit. Leaves slap and thrash. Tree swallows chitter high overhead, surfing on a sea of gas. They're black silhouettes against clouds the color of a bruise. Thunder grumbles closer. I count the seconds: two miles. A blue jay calls, unperturbed by the lowering sky.

The darkness deepens, and rain arrives in bursts. Yawp, my naughty crow, wheels past and tries to land on a foot-long branch at the top of the neighbor's spruce tree. Under his weight the skinny stick bends and points to the ground. Growling, Yawp climbs onto the upright leader, which accommodates him by bending into the horizontal plane. But the leader is bouncy, the wind is buffety, and Yawp flails. He soon quits and flies to the big silver maple. There the whole family gathers, then they swoop together into the midsection of the South Neighbor's tall pine. They work their way in to huddle near the trunk. The need for shelter has trumped the need for surveillance.

A flash glares over the trees, and five seconds later sound waves ripple across the yard. *Rumba-bumba-bumble.* One mile and closing. A tall, white fork stabs the planet west of me. A bolt slaps left to right, cloud to cloud. *Smack! Craaack!* There's little rain to dim the light show. Yawp, hungry, bored, or awed at Nature's spectacle, yawps tentatively from inside the pine tree. Then less tentatively. *Yawwwp . . . Yawp. Yawp! Yawp! Yawp!* Then comes a noise like a giant throwing a punch through a stack of windows.

The clouds, which now seem to brush my chimney, crash, snap, and roar around the crows, and me, and the songbirds who hop deep into their own trees and bushes. A solitary cardinal remains on the outer limbs of the chokecherry tree, fluttering and grooming in the thickening raindrops. Overhead, hooves gallop hither and yon. A spark splits into three prongs that skewer the earth. *Whhhack!* The cardinal retreats into the chokecherry, and the fire siren down the street commences wailing. For my part, I recollect my inten-

168

tion to refresh the old fire extinguisher. It's the kind of chore you only remember when Nature toys with the idea of burning your house.

The rain arrives, drilling into the ground and driving up the smell of damp dust. The world goes gray and water rushes down, hissing like a waterfall. The topography of the clouds is blotted out. Lightning becomes a muted flicker. Harder and harder the rain hammers. Behind the rain, thunder shudders like a freight train bearing down on the house. A huge energy churns up there. Lightning that strikes a mile away triggers a ripple of gas molecules that drums on my chest. A fire truck leaves the barn across the way, as fresh wailing rises from the downtown. Then, after just ten minutes of cataclysm, the rain calms. The flash and rumble move downwind, over the ocean.

The crows emerge from the spruce tree while sparse rain still falls. They fly into the tall trees and have a big talk. Then they glide down to North Neighbor's yard, where they hunker in the grass, bobbing and fluttering to splash water on their backs. They extend their wings and let the last raindrops pelt them. Yawp stands up and approaches a bathing adult. *Yawp* . . . He is pecked at. He bounds to his next victim. *Yawwwwwwp*. The air is thick. A fungus scholar once told me that rain causes fungus spore-pouches to burst and release kazillions of spores into the air, and that the smell of baby fungi is mistaken for the smell of clean air.

Thus come the summer rains. They're all we get, really, these flashy downpours. They're often insufficient to keep tomato plants from wilting and grass from going dormant. This is one of those years. The more substantial storms have migrated north for the summer, along with the jet stream that escorts them around the planet. We're left with the weekly thunderstorm, whose rain falls so fast the soil can't absorb it before it rolls away downhill.

🐿

"Maine: The way life should be." A sign bearing this motto greets visitors entering the state on the Turnpike. Some people might

169

quibble over the high dioxin content of lobster livers, or the acreage clear-cut by paper companies, but even those people would have to admit that most of the state smells the way life should smell. All the greenery helps. Pines especially release perfumey molecules into the air. The scarcity of heavy industry helps, too. But in my yard, the most delicious smell comes up from the ocean. The air in every corner of the planet smells a little different. My corner smells best.

Following the storm, the cool, northern air dominates for a while. It's flavored with the breath of a thousand miles of forest it filtered through, en route from Canada. I take out an atlas to see what else this air might have gathered.

Immediately upwind from my yard is Portland Harbor, where cargo and cruise ships belch diesel soot. Beyond the harbor, the smell of scorching coffee beans rises from the cafés of the Old Port. A pinch of coffee soot is probably drifting across my yard now. Continuing northwest I encounter the hometown of a paper mill, and the garbage incinerator that converts my trash to more soot, gas, and ash. Perhaps my old vinyl shower curtain is returning to me on the wind, as dioxin. Then it's out to farm country, where the air would have loaded up on pollens and soil dust, fragments of dry cow dung, fertilizer and pesticides, and the fragrance of table flowers and squash blossoms. When the cold air came across Sebago Lake, it might have gathered the glassy shells of diatoms from the shore. Beyond that, the hills begin, mounding up to the White Mountains in New Hampshire. There, the spruce and hard-woods threw in their two cents' worth of chemicals. (Or even four cents' worth. Some plants release extra defensive compounds when they're wind-buffeted.) The bare mountaintops surely gave a few grains of granite dust to the rushing air, along with a whiff of radon gas. The Great North Woods come next, offering mile after mile of tree molecules. And Canada beyond is just as green, clear up to the thinning stubble of the taiga, where mosses and dwarf willow tickle the belly of the wind. And now all that air is coasting through my neighborhood.

Unfortunately, the summer wind often arrives from a less pure quarter. In mid-August, a more typical pattern holds sway, with hot air coming out of the west. Down at the beach one morning I look out to Peaks Island, on whose shore the houses usually stand like neat boxes. The island is a gray mound, fuzzed with smog. Out over the ocean the smog forms a layer the hue of a tobacco stain. Back at home my radio issues an ozone warning. Particle counts are climbing, too.

This air wasn't cooked up in any forest. Much of it was dished out by coal-fired electricity plants in the Midwest, which leak sulfur and nitrogen, mercury and carbon. This waste evolved as it migrated from the midwestern smokestacks to my yard. The summer sun baked it into ozone and a damp stew of acidic particles. Now that it has reached my neighborhood, we're instructed to stay indoors and close the windows. My crows, and the squirrels, and the oak and apple trees don't have the option of hiding from the gunk. They breathe what they're given. Maine: The Way Ohio Should Be.

On the best days, though, the farmland and the forests of inland Maine heat and rise. This sets up a dynamic like water flowing toward a drain. The sticky air from my yard flows toward the inland drain. And to replace my departing air, cool air flows in from over the ocean. Sometimes it announces itself with wisps of sea fog that evaporate even as they limp down the street. Other times, the fruity air arrives with no visible herald. Occasionally it comes at night, pushing through the screens and filling the dark house. However it arrives, it always smells like tomato leaves and watermelon. And seaweed. And briny little life-forms. And mud. And summer. It's divine.

Investigating the smell of the ocean, I find that I'm inhaling a myriad of chemicals. The wind blows salt crystals off the sea surface and into the air. Seaweeds exhale methyl iodide, which may combat infection for them as it does for us. They also contribute fragrant methyl bromide and methyl chloride to the sea breeze. Algae egg cells release sex pheromones into the water to help algae sperm cells locate them, and one of those pheromones smells just like tomato leaves. But what causes the watermelon smell? I follow my nose to Philip Kraft, who dreams up molecules for the fragrance

171

powerhouse Givaudan, in Switzerland. "The watermelon scent of the ocean is to me still a real mystery," he e-mails in reply. He thinks it might be a combination of algae pheromones and ozone. If he knew what it was, he'd bottle it. He has experimented with combining melon and ozone notes to reproduce ocean, but success is evasive.

When the cold weather comes, the ocean smell fades. Many odor sources chill out in the winter. Evaporation of smelly liquids slows as the hemisphere cools. Deciduous plants, who churned out sweet jasmonate and grassy hexanyls in summer, die and desist. The sweet evergreens squelch their metabolism. Molds go still, and yeasts that convert fallen fruit into a vapor of airborne wine contract into dormancy. The off-gassing carcass of a mouse or seagull slows, then halts in mid-decomposition. This is one of the ways air announces a change of season: It carries less. Furthermore, even those smellies that don't freeze have a hard time getting around. Smells travel best in moist air, but cold air is usually dry. So when the winter wind blows into my yard, it brings little to stimulate the nose. It's that absence of odor, that cold emptiness, that smells like winter.

Winter comes, straight out of the north, bearing nothing but clear sunlight. A front blusters through at midday, rolling garbage cans, whipping the raspberry canes, and snapping twigs off the apple trees. A neighbor's ailing maple loses half its leaves. Widespread frost is forecast for overnight, and a radio weatherman urges listeners to pick the last zucchini.

❧

Suburban air carries its signature smells. It also conveys an abnormal quantity of light. The nightly bloom of photons saves me from tripping on cracks in the sidewalk and illuminates the bogeyman in the bushes. But it disorients other creatures.

On a cold, still night in the fall, I hear the night migrants slicing through the air. My mother is visiting, and it is she who picks out the sound from the background hum of streetlights, distant traffic, sighing pine needles, ticking oak leaves, dryer vents, humming plumbing, a hundred muted television programs, and a hundred more conversations. I'm sorry to say I don't think I would have, although now I'm probably destined to listen for it every autumn night for the rest of my life.

"Watch the moon," Mom says. "You might see one crossing." But the sound is ethereal enough for me. The avian voices form an invisible lace that stretches across the blackness. *Tew! Tew! Tew!* If a blind person wants to sense the allure of meteor showers, she could listen to the night migrants. The disembodied rain of bird-calls is eerie, and lovely. Blinded myself by the darkness, I can only imagine the tiny bodies hurtling south between heaven and earth.

Mixed with the *Tew! Tew!* is an occasional *Twak* and *Cheep* in a lower register. "The high ones may be warblers, but I really can't say," admits Mom, who normally knows birds by their song. "They all have completely different calls for migrating." The wide river of birds flows on, and on. Down below them, our earthbound bodies grow cold. When I go to bed, I leave a window open a crack and lie awake listening. *Tew! Tew!* Such a long night lies ahead of them.

My first thought is that the songbirds are migrating at night to avoid predatory hawks. But that turns out to be a minor consideration. A better reason is that they've got to eat, and they can't catch bugs at night. So they pack on the fat during daylight, then make headway at night. There are other reasons, too. All that flapping raises a bird's temperature, which can be dangerous. To cool off, they have to evaporate water from the air sacs in their respiratory system. As with sweating marathoners, that causes dehydration. Flying during the coldest hours reduces their need for cooling. And lastly, birds get better mileage per calorie after sunset. Cold air is less turbulent and more dense than sun-warmed air. Just like airplanes, birds are most efficient when they're slicing through a smooth, dense atmosphere.

Night migrants pay a modern toll on their night travel, however. Light pollution is such a recent obstacle that birds haven't adapted to it. One banner night in Kansas, for instance, between five thousand and ten thousand Lapland longspurs fell at the foot of a 420-foot tower. The annual tower toll for the entire United States is between four million and forty million birds. Scientists, intrigued by the piles of songbirds who die around lighted office buildings and communication towers on migration nights, have tried to understand what's going wrong. The studies indicate that the birds aren't attracted to a lighted tower, per se. Rather, if they fly into its cloud of diffuse light on a foggy or overcast night, their little brains short-circuit. They simply will not leave the circle of light. They flap, flutter, hover, and circle until they smash into an unlighted guy wire or a fellow bird or drop from exhaustion. Scientists have experimented with turning off the lights when birds are in this holding pattern. The birdbrains sort themselves out and fly away.

Cities and suburbs are aglow with these fatal attractions. I often walk with a neighbor and her dog in a park that holds a cell-phone tower. Why no pile of birds? I wonder. One reason could be that scavengers—coons, cats, rats, and other vermin—have learned that tower bases make excellent eateries, and they swallow the evidence of the massacre. Another reason is that a stream of migrants is only likely to flow into a light-trap when clouds or fog obscure their view of the stars. So some nights are safer than others.

I do find a bright spot (or a dark spot, really) in my research. A "Lights Out" campaign in Chicago seems to be reducing the carnage around office buildings whose managers darken them at night. Anecdotal evidence of success includes a report from one building's maintenance crew who no longer have to shovel piles of dead birds from their roof after a heavy migration night. Birders who walk in the downtown area also say they're counting fewer carcasses on the sidewalk. A similar effort in Toronto not only lobbies for darkened buildings, but also mobilizes volunteers to collect the thousands of downed birds. They've found that one in three survives its fascination with light and can be released.

The streetlights in my neighborhood are less hazardous to high-flying birds. They're more likely to mess with my own vision. And

174

to tell the truth, city lights don't blind people to the curiosities of the cosmos nearly as effectively as does a failure to look up.

When the stars fall from the sky on a hard-frozen November night, all the stray photons of Portland and South Portland can't combine to drown out the show. Wrapped in my granddad's fake-alpaca coat, I take a chair on the deck at three A.M. Gray clouds of my breath drift against a black velvet sky. *Whisss!* A fireball streaks over Neighbor Hugh's house. *Whisss!* Another follows it. From the corner of my eye I spy one over the South Neighbor's roof. *Whisssssssss!* A long one. They don't make a sound, but they should. They're so sudden, so glaring. And some people do report that meteorites make hissing or crackling sounds. But not I, not tonight. Like alien visitors, these fireballs defy local custom and self-destruct in silence.

They *are* alien visitors. Meteorites are bits of sand and pebbles shed by a comet and left in earth's path. As our orbiting planet plows into them, they burn in our atmosphere. Left in the air is a puff of comet smoke. If just the outer layers scorch away, the core of the micrometeorite may drop into my frosted grass. This happens with a regularity that I find stunning: Every square meter of the planet receives one speck of space dust each day, on average. Statistically, my yard ought to gain upward of a hundred space-dust motes per day. In the dark with my head tilted back, I open my mouth hoping something celestial might fall in. *Whisss-whisss!* Twin fires streak over the rooftops.

Even in these wee hours, my neighborhood is bright. The firebarn is washed with blue floodlight. Porch lights blaze yellow. A string of peach streetlights marches down each avenue. To the northwest the dirty glow of downtown Portland bulges above the trees. And still I can see the shooting stars, perhaps one every ten seconds. On other nights, cupping my eyes to block this streetlight and that bright bathroom window, I've seen the northern lights flickering in rose and neon green.

That said, while I'm seeing six shooting stars a minute, Mom back in her country house watches a blizzard of burning hailstones over her hayfield. While I watch a shower, she sees a storm.

175

I comfort myself with the notion that the city lights are filtering out all the weak and unfit meteorites from my view. Only the brightest pebbles show in the suburbs.

The fuzz of urban light does erase the night's fainter lights. I wouldn't want to have to count the moons of Jupiter from my backyard. But in my experience, the moon of earth and a few stars are visible from nearly anyplace on the planet. I've seen celestial wonders from cities all over the world. So I'm hooked. Just as I'll cock my ear to hear the night migrants for the rest of my life, I'm also compelled to check the sky every night, even from the heart of the world's brightest cities, to census the stars.

~

Maybe the reason I hear no meteor whistling is that urban air thrums with excess sound. When friends from the urban core come to visit, they stand in the yard and look around. "It's so quiet," they marvel. Then one of the neighbors fires up a lawn mower.

What I hear is partly a matter of what I tune in. When I lived in New York City, English sparrows were forever nattering in the fence, the shrubs, and the ginkgo trees. But for my ear to strain their calls from the rumble and bleat of traffic required concentration. And as with the fainter stars, some sounds are simply out of reach in the city. I remember watching Canada geese V over my brownstone in the fall. I strained the muscles of my cochlea and heard nothing but human din. Here, the geese honk into hearing before they flap into sight. Sometimes they've just risen from a nearby pond and are jockeying into position. Bits of V form, break loose, and re-form. Other times the V is high and precise and set on a course that will take the geese south over the ocean to Cape Cod.

As with light pollution, noise presents animals with a newfangled challenge. Birds especially rely on their voices to stake out a territory and attract a mate. Automobile traffic penetrates more bird territory every year. One bird, at least, has found a way to make himself heard above the roar. Dutch scientists set about

176

recording male great-tits singing near busy streets, and in quiet neighborhoods. When they analyzed the pitch at which the chickadee-like birds sang, they discovered that street-side birds were tightening their strings, so to speak. They sing higher to be heard over the bass roar of cars. Averaged over many individual birds, it's the equivalent of stepping from a white piano key up to a black one.

~

Our human landscapes change fast, faster than most animals can evolve. So an animal with flexibility built into his character is the one most likely to succeed in adapting to the changes in the air. Climate change, though, is going to test the stretch of even the most pliant animals and plants.

Climate change is nothing new. It's a constant on this planet. Sudden climate change is par for the course, too. Even a climate shift that coincides with a plague of extinctions is precedented. What's new about this upward bounce of the mercury is that we're causing it. The hardships confronting plants and animals are attributable to my Toyota and my electric oven. A second novelty in this climate wobble is that we know our planet-mates well enough to feel regret for the trials they'll face.

Different backyards are presenting different trials, which is why scientists use the term *global climate change*, not *global warming*. Northwest of my yard, the core of Canada has already warmed by seven or eight degrees. But easten Canada and the Maritime Provinces have cooled at a comparable rate. Here in Maine, the winter nights are warmer. Ice storms are replacing snowstorms, and the snowpack is melting one or two weeks earlier in springtime. Local sugar maples aren't getting the frosty spring nights they're accustomed to, so the maple-syrup industry is moving north. The ocean downhill from my yard has heated six degrees in a century. All these changes favor mobile and adaptable species and pose a challenge to those who need a specific habitat or a particular food in order to reproduce. Besides crows and squirrels, the survivors include a suite of aggressive invaders from other ecosystems. In my youth, deer ticks were unheard of around here. Now they're pressing deeper into the state every year. The milder winters are killing fewer mosquitoes. While that's good news for skeeter-eating bats and birds, it's not propitious for birds who are vulnerable to West Nile virus. The warming ocean may not kill crustaceans outright, but a depressing reality is that warm water makes bland lobsters.

Some animals will fade out, some will adapt. Migratory birds seem to be adjusting to an earlier spring. Nearly every species that migrates through Helgoland Island, north of the Dutch coast, now arrives earlier than four decades ago. Some come almost two weeks earlier. Similar "early bird" patterns are showing up elsewhere, too.

Whether the migrants are also heading south later this autumn than they used to has received less attention from scientists. The night I heard the songbirds *tewing* through the blackness, I noted the date. Next fall, maybe I'll be lucky enough to catch their voices ringing through the ether again, and I'll check the calendar to see if they stayed longer or shorter. Generally, it takes many years to accumulate a reliable quantity of data. But you have to start somewhere. And really, it'll just be an excuse. When winter is filtering into the air, and it's cozy indoors, the possibility of catching the night migrants will be a reminder to step outside and take the air.

WINTER

10

THE THIRTEEN COLDEST DAYS
OF THE YEAR

ONE OCTOBER MORNING, I look out the window and see rain pouring off the roof. A minute later I look again. The air is filled with feathers.

Snow? In late October? Around here, the first snow usually comes a month later. It typically melts upon landing and doesn't make a serious return until January. White Christmases, if memory serves, are a rarity. But this won't be a typical winter. This winter will be a bug-slayer. It will be a winter to shiver every scrap of fat off the breast of a chickadee. It will be a winter that finds many a bachelor squirrel begging entry at the door of a girls-only drey, a gift of bedding clenched between his chattering teeth.

In a couple of months, when my lawn's latitudes are tilted more steeply away from the sun, the snow will be made of fine crystals that baffle sound. But today's flakes are the size of Wheaties. Their crash landings in the grass make a sodden sizzle. With so many leaves still on the trees, it's an odd scene. Yawp and an adult crow are worming. Yawp hops into the air to snap at falling snowflakes. Then both crows retreat to the top of an apple tree. As their feathers dampen, they hop down a few branches, to shelter under the foliage. That luxury won't last much longer.

After an hour the snow reverts to rain, and the menace of winter washes away. By afternoon, the insects have recovered from the warning. Gnats gyrate beside the deck, and Babbette the spider's front-porch sister has caught a fly. So, what's up with this early snow? What indeed, replies weatherman Dave Santoro: It's nearly normal.

Dave is chief meteorologist for the local CBS affiliate, WGME 13. He's the sort of weather guy who raises his substantial eyebrows

181

in excitement when he identifies an advancing cold front. He leans into the camera to describe how condensing moisture releases heat to the atmosphere. Dave is, to quote Dave, a "met-Herb." That's a geek who's geeky over meteorology.

"This October is colder than normal," he tells me one frigid evening between broadcasts. "Sometimes the prevailing winds just establish themselves and don't move. The jet stream can set up in a way that won't change for . . . sometimes months!" This proves prophetic. The cold snap will turn into a cold pounding that lasts five months.

The snow picks up in late November, with a series of half-inch dustings. For tracking animals, these are ideal. The mouse who hops across the deck in the night leaves an impression of every toenail. A crow, on takeoff, carves a feather-fan in the snow. From a squirrel's pawprint I can guess the number of nickels in his pocket.

But a day or two after each snowfall, the air warms and the snow trickles away under the flattened grass and brown leaves. For too long, the ground is bare. Cold air seeps between grains of soil. Chilly rivulets of air trickle down wormholes and into the lairs of spiders. The cold sinks deeper into the earth every day. This is not supposed to happen. Sleeping roots and flower bulbs and worms and insects were counting on Nature to tuck them under a white quilt. Fresh snow is nearly as protective as fiberglass insulation, keeping in the earth's heat. Snow, if I had some, would slow the penetration of cold air into the soil. The earth might freeze only four feet deep, instead of five.

The winter solstice comes, but still no snow. My breath rises into a hard, starry sky when I say good-night to guests on the front steps. Finally, on Christmas, a storm plows across my yard. High above, water vapor has collected on circulating bits of desert dust, asteroid fragments, and forest-fire ash. As wind pushed the clouds east, these ice cores grew into starry flakes. When the flakes were too heavy to ride the wind, they fell. Now as they drop, a gale smashes them together. Their arms break off. Shards whisper against my storm windows. I can't see twenty feet to Neighbor

Hugh's house. West of here, in Albany, New York, the fragments are pouring down at five inches an hour. Even Texas is having a white Christmas. Under the heavy clouds, my yard is dark by midafternoon. And when we roll back into the sunlight on Boxing Day, the lawn is under an eighteen-inch blanket.

This is great for the subsurface folk. At last they've got some protection. But what will the abovegrounders eat? Until last night, the mice could doodle over bare ground in search of seeds. The white-crowned sparrows could pick over the dry grasses. The crows could hammer at frozen pears and snag hardy flies from the air. I go out to feed the crows. The dog kibbles sink out of sight, but the crows come down from their begging branch on the oak tree anyway. Chest-deep in snow, they stagger toward the food. Distracted by the fluff, they dip their heads and flutter, bathing. Squirrel Crook, whose tail bends in an L, hops past them, laboring to plow a path. The going will get easier in the coming days. The uppermost snowflakes will weld together, forming a crust that can support a crow or a squirrel. Then on New Year's Eve we'll get a storm much more typical to my yard:

Air warmed by the Atlantic Ocean swirls up the East Coast, dumping rain. It freezes on contact with my cold world. Every branch grows a coat of ice. The deck railing is so slick that starlings come in for a landing and slide right off the other side. They stumble like clowns. Out on the crusting snow, my crows skid sideways into little hollows. The young pines bend their needles to the ground. Up the coast a ways, people lose electricity. Pine branches, their needles clotted with ice, snap and gore roofs. Cars pirouette into telephone poles. Yes, this is winter on the coast of Maine.

Dave Santoro says the warm Gulf Stream out in the Atlantic contrasts with the cold continent and drives these sloppy storms up the eastern seaboard. I prefer the dry, cold storms that whistle in from the north. I imagine my animals do, too.

183

Why doesn't every living squirrel and crow either hibernate or hightail it south to avoid this mess? Why do they stick it out, wide-awake, through the blistering cold?

Well, I'm a potentially migratory mammal. Why don't I go to Florida? I suppose the practical answer is that going south would entail the expense of moving myself and then securing a southern territory. And when I returned in the spring, I might find my old territory had been invaded. Furthermore, I've figured out how to find food here in the winter, and I've adopted behaviors that reduce my risk of freezing solid. And that must be what a crow or a squirrel would reply, too. He would tally his expenses in body fat, rather than money, but the economic fundamentals are the same. So the answer must be that, for a squirrel or a crow, living outdoors in the screaming wind and snow is easier than migrating or hibernating. It doesn't look easy, though.

"The thirteen coldest days of the year, on average?" Dave Santoro queries from my TV screen on January 13. "When do they start?" His eyebrows leap. A blue word shivers larger and larger until it fills the screen: *TOMORROW*. On recent nights, the mercury has fallen to ten, even five degrees. Dave says the next few nights will go well below zero. When I strap on snowshoes the following morning, it's twelve degrees. I take a turn around the yard to see who survived.

Some mammals have the option of hibernating through the thirteen coldest days of the year. But most species have evolved to face them head-on. This morning, mouse tracks trickle down the deck steps and along the shoveled path toward the driveway. They turn and enter a hole in the snowbank. Kneeling, I find a series of tunnel doorways. I've been scattering seed on the path for ground-feeding birds. Who knows how the little rodents detected the seed, or how far they tunneled to reach it.

I felt sorry for the mice when the Christmas storm dumped eighteen inches of snow on them. But that snow was probably a blessing. For one thing, they now travel out of sight of owls, cats, and other nocturnal hunters. They can circulate from the bird-seed, to the stash of stolen corn, to a decaying jack-o'-lantern, all undercover. And they're probably warmer now, too. With a shovel I chop through eighteen inches of layered ice and snow, lay a thermometer on the frozen grass, and refill the hole. When I retrieve it a half hour later it reads twenty-five degrees—thirteen warmer than the air.

At this time of day, my mice are deep under the snow saving their fat. To make it last longer, they zipped up their coats back in autumn. The winter fur of a deer mouse is one-third warmer than the summer fur, and it's fluffable. The key to Owens Corning insulation is that the glass fibers trap air, and that slows the migration of heat. Mouse fur is like adjustable Owens Corning. On a day like this, my mice can fluff air into their fur. Then during the January thaw, should we be blessed with one, they can flatten their fur and squeeze out the air. The mice are also wearing long underwear. They probably began laying on brown fat when the days got shorter back in the fall. Like white fat, brown fat holds in body heat. But it can also generate warmth. In the summer, a chilly mouse might shiver, letting metabolism in her muscle cells raise her temperature. But that's akin to running your car's engine in a parking lot, just to keep the heater blasting. Brown fat produces heat all by itself, with no wear and tear on the muscles.

Inevitably, heat will radiate out through my hidden mice. But that heat needn't rush into the cold air. Under the snow and in tree cavities, the mice have built nests lined with plant down and shredded leaves. Like a sleeping bag, a nest arrests the flight of body heat. My mice can stretch their blubber even further if they all pile into a single sleeping bag. Scrappy and territorial in summer, mice in winter turn cuddly. As many as a dozen strangers will share a nest. With most of each mouse's surface packed against another mouse, there's less surface exposed to the air. They all stay warm longer. Huddling does have a serious drawback. It increases the amount of competition for food in the area near the nest. But the benefits of cuddling must outweigh the penalties. Huddling mice are more likely to lose weight than loners, indicating that winning food is less important to them than keeping the peace with their bedfellows.

185

Even with all these precautions, a mouse may still fail to scrounge up as many calories as she burns each day. In that case, she can lower her body temperature a few degrees, falling into shallow torpor. And it works. The mouse who spends this crystalline day cuddling with strangers in a nest, and who allows herself to go a bit sluggish, will burn less than half the calories required by a wakeful mouse who has neither sleeping bag nor friends.

Squirrels walk a finer line through winter. Stumpy, Crook, and friends time their mating marathon to coincide with the weeks when both temperature and food supplies are grinding along the bottom of their cycles. At my nearest apple tree, I examine a patch of snow on the twisted trunk. It's an amber color. Pee? I'm surprised to find the odor floral. But it must be pee. Circling the trunk, I discover that squirrel teeth have chopped chunks of bark from around a hole in the trunk. A male squirrel has left a message for the girls to read.

They've all been acting funny lately. I've caught Crook getting squirrelly with the shrubbery. A normal squirrel one minute, the next he'll ambush a nearby branch and whirl around it a dozen times. I can't tell if he's fighting with the shrub, or mating with it, or just blowing off squirrel steam. He might thrash for half a minute before popping off the branch and settling down. Without knowing the function of this behavior, it's hard to say he's wasting calories. He may be training for the Great and Gory Mating Race. That event will definitely be a caloric catastrophe, but it's not a waste. It could produce a new generation of little Crooks.

As the cold days pass, the prerace formalities intensify. One day Stumpy, my little amputee, is just another squirrel unearthing blackened acorns from the bare earth at the base of a tree. The next day, she's a hot ticket. Everywhere she goes at least one male follows. Usually two or three lope in her wake. And this is not the main event. The boys are just monitoring Stumpy's hormonal status. They don't want to be sleeping in on that morning when

186

she wakes up in a mating mood. So for a few days they forgo calorie-gathering and dedicate themselves to sniffing Stumpy's pheromones. They burn stored fat recklessly. Every trip up a tree trunk in pursuit of the girl torches another clump of calories. Every chase of a competing male is an expenditure. And Stumpy isn't the boys' only quarry. Every morning each male in the neighborhood is probably making a circuit to sniff half a dozen females.

When Stumpy's big day dawns, she pokes her head out of the oak-leaf drey she shares with her female kin. She's nose to nose with every boy in the neighborhood. They know. Some may already be bleeding from the ear or rump, having battled for a position near the nest. One may even try to crawl into the drey. She'll scold him out. And then she'll bolt.

The boys follow. Scientists call this event a mating bout. For squirrels, mating is a contact sport. As Stumpy zigs and zags, whirls up and down tree trunks, a pack of males jostles and snarls behind her. The forerunners expose their haunches and tails to the followers. The followers slash those haunches and bite those tails. The competition between males may get so frenzied that the boys lose sight of the girl. In that case, they may rush after a bird or any other squirrel-size distraction. Squirrel researchers Michael Steele and John Koprowski once rolled a softball through a mating chase and saw six males peel off the pack to court the ball. A male may eventually maneuver Stumpy onto a high tree branch and grip her for mating. But even if he can repel the rammings and slashings of the other boys, he could still lose the insemination privilege. Stumpy may jerk from his grasp in what's called a breakaway. She'll get herself down from the tree and into a safer, quieter spot. There she'll mate with the first male who finds her. This is the moment the younger men have been waiting for. Instead of wasting energy fighting with older males, some clever studs have been killing time on the ground, hoping for a breakaway.

The male who mates with Stumpy takes a couple of steps to insure the success of his sperm. To seal the deal, his ejaculate will harden into a "copulatory plug," which blocks her vagina. Then he'll guard her for a while to give his little swimmers a jump on

187

the competition. And there will be competition. Stumpy will groom herself, pull the plug, and resume the Mating Race. When the excitement dims with the afternoon sun, Stumpy will have collected sperm from two to four males, some of them very fit, some of them very lucky. Some may have given the race every-thing they've got. In *North American Tree Squirrels,* Steele and Koprowski report watching a fox squirrel crawl out of bed one morning and tumble dead to the ground. The rogue had mated with five females in the previous days, and an autopsy showed that he starved to death. Among my local boys, I notice only minor damage after the mêlée. All sport tooth-size gaps in their tail fur, and a couple have torn ears.

As for Stumpy, she faces the prospect of growing squirrelettes in the dead of winter with her food entombed under eighteen inches of snow and ice. But there is a method to this midwinter madness. The arrival of the kids will coincide with a springtime flush of food. And based on the number of squirrels in my yard, they're managing fine. Nonetheless, I feel sorry for the girl. I take to scattering a corn-and-sunflower mix on the snow every morning. Crook and Fluff, my big-tailed squirrel, eat most of it. But Stumpy comes around some afternoons for the leftovers. If I step out to toss her more, she wheels on her peg leg and dashes for the shrubs.

Most of my mammals trudge straight through the winter along with the mice and squirrels. Shrews hunt chilled insects. Possums mine compost piles and gestate possumlets. Even Stunky the skunk releases the occasional cloud of chemistry from under the porch, although I can't find a skunk track to save my life. But one animal has dropped completely out of sight.

Crunching past the shed, I check the woodchuck hole at the corner. I stand in the glassy wind and listen, knowing I'll hear nothing. If Big Fat Momma is directly beneath my snowshoes, I'll never know it. Even if I were inside her hibernaculum, there would be little to hear. Wherever she curled up, she's inhaling only every few minutes. If I had a stethoscope pressed to her chest, I'd have to wait fifteen seconds to hear a thud.

Hibernation is a rare art. Everything I thought I knew about it is wrong. Black bears, for instance, the poster critters for hibernation? They don't, technically, hibernate. They inhale too often, they stay too warm, and they wake up too easily. They don't qualify. Animals who zonk out for a few days in a row—such as skunks, raccoons, and chipmunks—aren't hibernating, either. But the woodchuck: That's an animal who can hibernate.

When Big Fat Momma entered her grass-lined hibernaculum last fall, she reduced the rate at which her cells burn fuel. The cool-burning cells needed less fuel delivered, so her heart rate drifted down to a few pumps per minute. Her cells called for less oxygen, so her respiratory rate dwindled to a breath every three or four minutes. (By contrast, a chipmunk who sleeps off a few days of hard weather still breathes twenty times a minute.) Big Fat Momma's slow metabolism produces less heat, so her body has cooled to just eight degrees above freezing. To avoid icing up, she's burning a small amount of brown fat.

If that were all she had to do, hibernation would be a snap. But there are problems. One is that brown fat requires water to burn. Under the snow, under the ground, Big Fat Momma's body is becoming dehydrated. This problem is eased by another problem: She's burning up muscle, too. That will leave her weak in the spring, but at least the process returns some water to her blood. The second difficulty is that she has to warm herself and arise from hibernation about every two weeks. This is hideously expensive, in terms of stored fat. Hibernators blow at least half of their winter blubber in periodic heat-ups.

Each time Big Fat Momma wakes, she'll drag herself into a latrine chamber to defecate. But that may not be the reason she spends so much fat to wake up. It may be that she desperately needs sleep. Strange as it sounds, brain waves in western ground squirrels reveal that they start running a sleep deficit when they hibernate. The deficit builds. After a week or so, a hibernating

189

squirrel pours a bucket of fat on the fire and raises his body temperature to the normal range. And then he sleeps a normal sleep. After about half a day, he throttles back and returns to deep hibernation.

Rousing may serve a combination of purposes. A young psychologist from Ohio State University suspects hibernators wake up to give their immune systems a chance to hunt down germs. Brian Prendergast injected hibernating golden-mantled ground squirrels with bacterial fragments that should have incited their immune systems to spike a defensive fever. Nothing happened. At least, not until the squirrels naturally cycled into a warm-up a few days later. Then they spiked the fever. Prendergast theorizes that white blood cells can't function at low hibernation temperature. Through the winter, Prendergast found that these heat-ups consumed up to eighty percent of the squirrels' fat, so whatever they're waking up for, it's important.

Big Momma breathes on under my snowshoes. It isn't the easy life it's reputed to be. Whereas the shrews who hunt all winter can gain weight in the cold, Big Fat Momma is wasting away. But none of the paths through winter is easy. The woodchuck hibernates because she eats only green things, which dried and died last fall. She's hoping that if she burns calories slowly enough, she can make her dinner last until breakfast is served. In contrast, those squirrels and mice who invested their energy storing food for the winter now risk having it stolen, which would be calamitous. Even the raccoons and skunks, whose dietary flexibility allows them to eat whatever they can dig out of a snowbank, risk being interrupted by a dog or coyote as they hunt calories. Winter is just hard.

On frigid days like this, small birds are rare in the yard. Perhaps they're perched outside the travel agency window, dreaming of Costa Rica. The crows, though, never miss a day.

These mornings, they come home from the rookery hungry.

November and bitter December they stayed, scratching at the lawn. When the snow flew on Christmas, I thought for sure they'd scram. They didn't. Shrugging off the English sparrows' bullying, they set up house in the forsythia hedge and collected the seeds I scattered each day. I consulted my *Peterson* guide for a map of the bird's winter range. The blue shading ended about three hundred miles south, in Long Island, New York. But my guide was published in 1980. On the Internet I tapped up a newer map and found that the blue now washes all the way up the coast of Maine.

Whether the white-crowned sparrows are adapting to northern living or are taking advantage of climate change, they chose a rough year to push their limits. The atmospheric gears that pull the jet stream across the Atlantic are sluggish this year, according to Dave Santoro. As a result, the jet stream, which marks the boundary between Canadian frigid air and continental pretty-cold air, has sagged deep into the Southeast. The eastern United States is being blitzed with blizzards and harsh freezes. One windy night the hard air penetrates the northwest wall of my house and freezes the water inside a valve of the dishwasher. When the appliance thaws in the morning, a flood sweeps across the kitchen floor. Wringing out towels, I contemplate anew the economics of migration. But the saggy jet stream doesn't impel the white-crowned sparrows to run for Long Island. Perhaps they can't afford to stop eating long enough to pack their bags.

Getting through a winter day isn't easy for any bird. And the nights can be fatal. But of course, they have their tactics. Come nightfall, the songbirds find a sheltered spot in an evergreen. There they probably shiver all night, producing heat in their moving muscles. They fluff air into their feathers. They minimize their surface area by tucking head against back. Some even drop their body temperature a few degrees and become slightly torpid. I suppose it's possible some species will huddle like mice, although this has been documented only once, in a forest-dwelling kinglet. Even with all these precautions, though, most birds carry only enough fat to get them through one brutal night. Come daylight, they have no choice but to find more fuel.

My reaction to this knowledge is predictably unobjective. Now that I know the peril my birds face, I feed them faithfully. The cardinals get sunflower seed. The sparrows and juncos get small seed. So my songbirds won't weary their little legs, I pack the snow for them. The crows get their kibbles and table scraps. And in a voice that seeks the middle ground between attracting birds and confirming the neighbors' worst suspicions, I whisper-yell, "Crow-crow! Crow-crow!" And they come. The crows love meat best. Nuts are also coveted. Dry cat food is acceptable, but perhaps a challenge to swallow—they spit out the X-shapes over and over, until they find a comfortable fit. Small-kibble dog food is better. Crackers aren't always worth the bother. Bread is okay, but large pieces of any food attract seagulls.

The *whuff-whuff* of seagull wings blows a cold wind of hypocrisy in my face. How is it that I can fret over the cold toes of the crows, then pound on the window when a seagull or a starling descends? The starlings, to speak in my own defense, travel in packs that can vacuum up a pint of kibbles in sixty seconds, while the crows are still tiptoeing across the snow watching for alligators. And starlings are a European species that needs no help conquering the world. The gulls . . . well, they're just as greedy, and six times bigger. The uncomfortable thing is that they're native birds, holding a valid passport to this neighborhood. Why not feed them? When I ponder this, I recall that I only started feeding the crows to lure them into the yard so I could watch their fascinating behavior. And seagulls aren't fascinating. I grew up with them and never saw one commit an interesting act. Even the one my sister raised from an egg impressed me only by how large an object he could swallow. So with almost no guilt, I go on throwing rocks, garden clogs, and mittens at starlings and gulls.

❧

It's late January when I remember that the little peach tree is standing unprotected in the yard. Its tender bark is exposed to mice, who will tunnel to it and strip it. With its circulatory system girdled, the sapling will emerge from winter as a dry stick. I meant to wrap it before snow flew, but I meant to do a lot of things before snow flew. I head out to trample the snow around the trunk. If the mice want the peachling, they'll have to come out on top of the snow and do their damage in broad starlight. In my trampling, I nick the trunk and am surprised to see electric green beneath the bark. But it goes to show: Even those who hibernate through the frosty months don't shut down completely. Most organisms have to stay a little bit awake to stay alive.

Many plants, like dandelions and milkweed, don't even attempt to survive. These annuals throw all their energy into seeds that will revive the family name in the spring. Then they fall down dead. But perennial plants, including my oak tree creaking in the wind, have to stand and confront the cold. Unlike animals, plants are sessile—nailed to the spot—which rules out hibernation and shelter-building.

Once again, my plants demonstrate more smarts than I would have given them credit for. Back in August, the oak tree and the pines noticed the days were getting short. A hormone that halts growth suffused their branches and leaves. Each cell, anticipating that ice crystals would form around it, hardened its linings to prevent puncture. The trees, and all the other overwintering plants in the yard, assembled freeze-resistant molecules to stir into their sap. Some replaced water with sugars. Some adjusted their proteins. Some added fats to their needles. Some mixed and matched strategies. However they did it, by the end of August my trees were ready to fight a freeze. They rested, and waited. Frost hit on October 9. All around the trees, annual plants went rigid. Dew that settled on the tomato leaves froze, then a crystal contagion spread inside the leaf tissue itself, gouging cells. When the morning sun struck the frosted leaves, the crystals thawed, the cells bled, and the leaves went limp. Standing tall amid the ruin, my trees got back to work. Through October, they laid in more antifreeze every day.

The mercury drifted down, and my oak tree's twigs chilled along with the air. The day the twigs fell a few degrees below freezing, they got a quick fever. Is my oak tree warm-blooded?

Given the ability of cold-blooded insects to warm themselves, and warm-blooded birds and mammals to chill themselves, one could be forgiven for asking. But this heat burst is the signature of ice. Crystals have just formed in the spaces between the tree's cells, and that shift from liquid to solid released heat. It was a fleeting reprieve. The tree reaped a more lasting benefit from the ice itself, which pulled water from inside the cells. The cells were left holding a syrup of antifreeze. By the time the sun had resumed its winter habit of setting at noon behind the South Neighbor's roof, the tree was nearly finished. Each leaf had shunted unused nutrients back into the branches. A layer of cells grew between leaf and limb, cutting off circulation. The green chlorophyll faded from the dying leaves. And in mid-November, a big wind knocked down all the dead leaves in one night. The sumacs had lost their leaves many weeks before. So had the apples. The lilacs and the chokecherry were bare. My little forest was gray, and prepared.

Except for the pines. It's hard to gauge the preparedness of a pine, because it hangs on to its needles. It seems like a costly thing to do. Each leaf can leak moisture to the dry winter air, dehydrating the tree. And each leaf has to be saturated with antifreeze. So why not chuck 'em, like the deciduous trees? Again I turn to *Life in the Cold*. The leaky-needles hazard, Peter Marchand contends, is a bit overblown. Evergreens seal their leaves with a high-performance wax that works especially well in the cold. And evergreen stomates, the little air-ports on the underside of a leaf, can seal much tighter than deciduous stomates. I'm surprised to learn that the worst hazard winter can present to my pine trees is a sunny day. Then, water vapor trapped in the dark needles heats up and wants to escape into the cold air. Because of this, my pines will dry out a bit as the winter progresses. Even so, one of Marchand's own experiments demonstrates that if you snip the trunk of a spruce tree under the snow, the tree can stay hydrated right through winter. Only when spring temperatures heat the truncated tree will it be forced to exhale fatal quantities of moisture.

So, despite their need to winterize every needle, the pines know

196

what they're doing. They're going to save a bundle of energy next spring by reusing last year's leaves. And on the day the sun climbs over the neighbor's roof once more, the pines will be ready to collect it and grow. They may be back in the light-harvesting business as early as February.

In a natural forest, pines and oaks compete for sunlight, with the prize going to the tree who grabs enough of it to grow tall and shade out the competition. If the pines have evolved a way to collect light earlier in the year, the oaks had better do likewise. That could explain why the stem of the little peach tree was green under the bark. Peaches, oaks, and other deciduous trees keep chlorophyll in their bark through the winter. When the pine needles start hauling in the sunbeams in February, my oaks may begin gathering it, too, with their leafless twigs. Dead as my forest looks now, even buried branches may be churning out a few carbohydrates, by photosynthesizing light that penetrates the snow. As the snow ages and turns icy, still more light will penetrate, waking the buried plants. My trees may even be drawing a small drink of water out of the frozen yard. Experiments show that water does rise slowly through the woody trunks of various trees and shrubs in midwinter, even when the plumbing under their bark is frozen. Where they find this beverage is problematic. Peter Marchand specu-lates that even in frozen ground, thin films of liquid water may coat some soil particles.

Beneath my boots, beneath the soil, the roots of all the perennials are whistling in the dark. They're not dead, and they're not frozen. The oak roots and the bamboo roots, the tulip bulbs and the iris tubers, all weatherproofed themselves in the darkening days of fall. They cast off their green uppers, and withdrew under the protection of the soil. The earth turned rigid around them, but the antifreeze in their cells buffered them from the cold. Like the woodchuck, they wait.

I take shelter from the wind in the lee of the far apple tree. Even at high noon these days, the sun doesn't get halfway to overhead, but it does deliver a little heat. The tree's dark trunk is absorbing some of those rays. I poke around the flakes of bark, looking for insects who might be thawing. My mittens are too bulky to explore with any grace, and it's too cold to take them off. But I know the insects are here. Only one, the monarch butterfly, is truly migratory. So the rest must be here. They must be everywhere. Stiff larvae, shiny pupae, clusters of eggs, even adult animals, are stashed under bark, under stones and fallen leaves, and scattered through the earth between the shelters of mice and woodchucks.

Insects lack the insulation that birds and mammals rely on. Naked and tiny, they stare down snowflakes larger than themselves. Many take the approach that annual plants do: They lay indestructible eggs while it's still warm, then give up the ghost. But those who brave the winter undergo transformations that belie the insects' reputation for simplicity.

Of all the overwintering geniuses, the goldenrod gallfly is extratalented. This walnut-colored fruit fly lays an egg on a shoot of goldenrod, a tall annual plant with an August spray of mustard-colored flowers. The gallfly's worm gnaws into the shoot and, in a flourish of genetic engineering, causes the plant to grow a one-inch bead of flesh on its stem. Inside the gall, the larva eats goldenrod tissue, molting twice as she grows. But in her third "instar" (incarnation, roughly translated), she senses a chill in the air. She modifies her own innards now.

As the days shorten, the larva's body manufactures granules of calcium phosphate, the stuff bones are made of. She also brews a slew of sugars and alcohols that lower the freezing point of her body fluid. The mercury ticks downward while she eats on, until the cold halts her chewers. One night, frost on the goldenrod stem inoculates the plant cells with ice, and the plant dies. On the brown stalk, inside the gall, the larva lies motionless. The following weeks shove the temperature in the gall down to thirty degrees. The larva remains soft. At twenty degrees she's still soft. And at fifteen degrees. She's supercooled: below freezing, but still liquid.

Most insects can do no more than this to fend off ice. They

198

don't want to freeze, and so they perfect their supercooling. For instance, the flesh fly (that's the dog-doo-eating, carrion-colonizing look-alike of the housefly) can stay crystal-free to ten below zero. Supercoolers usually rely on protections beyond their antifreeze. They build cocoons or find sheltered spots, either under a few inches of soil, under bark, or in the eaves of my house. And whereas the goldenrod gallfly manufactures little particles in her body, serious supercoolers purge themselves of anything that might give ice a foothold. They empty their gut to be sure no bacteria or fungi remain for supercooled liquid to freeze upon. They cleanse their blood of food particles that could give ice a grip. Heaven forbid someone should drop a snowflake on a supercooled insect— it takes very little to cause a storm of crystallization in such cold fluid. But as long as a supercooled critter doesn't ingest, inhale, or bump against a particle that could set off a chain reaction, he'll be fine. Down to a point. (For a yellow-jacket wasp, the point at which supercooling fails and crystallization kills is nineteen degrees. The record for superest cool is held by a beetle of the Canadian arctic: seventy-eight below zero.)

The goldenrod gallfly, however, is protected even from cata-strophic crystallization. Waaaay back up the thermometer, the larva saw a supercold day coming. Instead of purging her particles, she deployed them into her body fluid. The circulating calcium phosphate specks invited her liquids to freeze, and they extended the offer so early that the crystallization was gradual, not cataclysmic. In the spaces between her cells, the fluids solidified. Water inside the cells seeped out to join the growing crystals. Antifreeze left in the cells strengthened her cell membranes to prevent collapse. Her cells, now holding syrup, will not freeze. Well, not unless the mercury hits forty or fifty below zero. Then her goose will be cooked.

If it's a quick hit of cold, perhaps she'll only be injured. Flesh-fly pupae that get too cold may live through the freeze, only to be stuck in their cases. If they can emerge, nerve damage may prevent them from eating, or grooming themselves. They may not be able to reproduce. They may wish they hadn't survived. And a deeper freeze is, of course, completely lethal.

199

Insects employ their antifreeze strategies at all stages of development: eggs, larvae, pupae, adults. Whichever phase faces the winter battens the hatches and hunkers down. And while insects' abilities are remarkable, they're not foolproof. A cold winter, especially if there's no insulating snow, plunges icicles into the hearts of a huge percentage of insects and their eggs. Even a soft winter can be hard for some. The goldenrod gallfly will burn too much energy if the cold doesn't put a halt to her metabolism, and then she'll likely die. She may die even in perfectly cold weather, because chickadees and woodpeckers have learned to recognize those brown bulbs on goldenrod stems. Like children cracking walnuts, the birds breach the pithy walls and swallow the carefully cooled meat.

Today, I can detect no insect motion in the yard. But under the snow I know some exceptional animals are working straight through the winter. Investigators with more foresight than I have sunk pitfall traps into the earth before it froze and have been rewarded for their effort. From beneath the snow on a frozen lawn, they interrupted the winter work of ants, springtails, mites, and nematodes.

January delivers its thirteen cold days, and then some. February will prove just as hoary. This winter will win awards for the length and depth of its coldness. It'll shape up to be an excellent winter for stabbing insects. The spring will deliver a smaller crop of ticks, mosquitoes, and termites. But there will also be fewer butterflies, ants, and robber flies. And my birds, with hungry families to feed, will notice the difference.

~

At this time of year, the north pole leans away from the sun. During the day, the sun's rays strike people at my latitude in the ribs. The top of our heads go unlit. The sun sets shortly after four in the afternoon. And at midnight tonight, the trees in my yard will be pointing almost straight away from the sun. That's how far the planet tilts.

We plants and animals of the northern latitudes spend long winter nights facing the darkness of space. Tonight, when Indonesia is being broiled, the occasional photon will scatter around the globe to zing through our sky. But photons will be few and far between. Winter up here on the shoulder of the planet is just plain dark. It's cold. It lasts a long time. And we who stay in our home latitude are hard-pressed to get through it without taking a bite out of a huddle-mate. Even Dave Santoro, who I thought could never be persuaded to speak ill of the weather, is running out of patience with the cold.

He says, "Who landed in Jamestown and said, 'No, no, no. This isn't cold enough. I'm going a few hundred miles north where the winters are longer'?" He throws up his hands and lets them fall. Outside my window, the trees stand motionless in the dark, their heads like dull arrows pointing toward another long night.

11

MELTING POT BLUES

THE LEFTOVERS OF my sharp-shinned hawk's lunch lie at the center of a circle of feathers. The lunch's breastbone is blue-white, bitten ragged by the sharpie's bill. Maroon flesh clings to it. The ribs are intact, but the gut is gone. The gizzard dangles at the end of the esophagus. The heart glistens in the chest cavity. The thigh meat has been stripped away, and the skin of the back is missing. Bits of yellow fat remain where the breast meat was. The wings spread wide, and the feathers are iridescent green. The head is thrown back in the oak leaves, too, exposing a black velvet throat and a yellow bill. Half an hour ago when I saw the hawk wrestle its meal to the ground, I couldn't tell who that meal was. Now I can see it's a starling. Couldn't have happened to a better bird.

As a botanist friend told me this summer, my yard typifies a problem that's global in scale: Foreign plants and animals gain a foothold on the shore, then drive the natives into the deepest woods and harshest deserts. In my yard, the conquest is nearly complete. Among the plants, outsiders outnumber natives four to one. For mammals, it's a closer contest, with foreign rats and cats facing off against native raccoons, skunks, possums, mice, and shrews. As for the birds, I'd have to say that the foreigners are winning. The forsythia hedge is lousy with English sparrows. Pigeons descend in a mob that scatters the local juncos and cardinals like leaves. Starlings are so numerous that they swoop over the yard in throbbing clouds. Well, they're one starling less numerous today.

How do they get to this continent, these not-so-huddled masses? The starling is unusual: He was brought intentionally, lovingly. Most invaders come by accident. The Norway rat stowed away in the holds of ships. The earthworms who are altering northeastern forests probably came in potting soil, ballast dirt, or along with dirty livestock. More recently, the Asian long-horned beetle entered undetected in lumber from China. But the starling was introduced on purpose. In 1890 human immigrants in New York City imported and freed sixty of the birds in Central Park. The members of the American Acclimatization Society had taken the notion that they themselves would acclimatize better to their new homeland if every bird mentioned in the plays of Shakespeare was here. Many of the transported starlings perished. The survivors founded a population that today sweeps from coast to coast. (Northern Canada is too cold for them, limiting their conquest of that realm.)

The result has been less poetic than its inspiration. Starlings destroy crops. Their noisy hordes cause other birds to flee. In breeding season they wrest nesting cavities from woodpeckers, including the rare redheaded woodpecker. And when starlings congregate in roosts, their droppings become a breeding ground for fungi that cause blastomycosis and histoplasmosis in people. Ecologists used to think that an invading species required an open niche in the ecosystem. Starlings demonstrate that a good invader creates its own luck, by taking someone else's niche. In my yard, starlings not only monopolize the best tree holes, they even steal food from the crows. When I toss out kibbles, a flock of starlings often touches down like a tornado, snatching up the food while the crows are still double-checking the lawn for hidden perils.

Maddening as starlings are, if I were given a vote about whom the hawks should kill in my yard, I'd request that they eat the English sparrows first. Starlings may quash the woodpecker population by denying them places to reproduce, but at least they don't kill other birds outright. The English sparrow is a murderer. He pecks the heads off other birds and, through West Nile virus, may even be picking off people.

In my yard, the English sparrows monopolize the best winter

habitat, the dense forsythia hedge. When they're not squabbling amongst themselves, they issue a loud, monotonous *"Cheat! Cheat! Cheat!"* They're not even nice to look at. The males have a gray cap, a chestnut back, white cheeks, and a black bib. The females are dull beyond description. The beak is thick, confirming that this sparrow is actually a finch.

I'm giving up on scattering regular birdseed, because of them. They flock to it, then bully any hungry natives who dare to come forward. Juncos are driven off, as are white-crowns. Cardinals, being twice as big as a sparrow, are slower to retreat, but the sparrows are persistent. Back in berry season, I watched the sparrows target a cardinal who was sitting on the fence near the raspberry canes. A handful of sparrows flew to sit on either side of the cardinal. They jiggled and jostled, harassing the cardinal with bodily proximity. They hopped over the cardinal to sit on his other side. They yapped and cheeped. Finally the cardinal gave up, flying away from the food. Another day I watched the sparrows hassle a downy woodpecker, following her from one apple tree to the next, surrounding her and yelling wherever she tried to feed. She could ignore them for a few minutes, then she'd wing away for a break. Each time she returned, they mobbed her anew. I grow to hate them. I dream of mist-netting them and twisting their little necks, slinging them into the Bamboo Wilderness for the skunks.

I'm relieved to learn I'm not alone. I had heard gossip about English sparrows killing bluebirds. When I went looking for confirmation, I located my fellow sparrow-haters. Their trapping inventions and killing techniques are easily located on the Web. The sites are illustrated with candid-camera shots of sparrows pecking other birds' eggs, and of baby bluebirds tossed from the nest to die on the ground. I can imagine the rage this inspires in the breast of a bluebird lover. Eastern bluebird numbers have plummeted as sprawl and farms consume their habitat. Aiding their survival is an impassioned network of people who erect nest boxes for them. Starlings can be excluded by limiting the size of the doorway. But English sparrows can squeeze through a smaller hole. And even if an English sparrow doesn't want a bluebird's box for his own brood, he may still enter and kill the bluebird, on principle.

And so I discover a world of sparrow-scaring devices, sparrow seed-traps, and sparrow nest-traps. As for dispatching bagged sparrows, the list of methods is long: Compress the chest for thirty seconds. Grip the head and crack the bird like a whip. Swing the sparrow-in-a-bag against a solid object. Submerse the bag in water. Step on the bag. A bird researcher I once worked with taught me the chest-squeezing method, but I suppose he wanted his dead birds to look nice. Quicker is kinder, to my mind. I guess I'd opt for the solid object. Because the sparrows aren't native, it's perfectly legal to do them in. The same goes for pigeons and starlings.

With West Nile loose upon the land, the English sparrow has *cheated* even higher on my list of enemies. The sparrow evolved in Eurasia, overlapping with the virus. His immune system can fight the bug, and he carries loads of the virus around in his blood for mosquitoes to tap into. Whomever a mosquito subsequently bites gets both a welt and an injection of West Nile. My crows are terribly susceptible to the virus, and here they are sharing a yard with these germy English sparrows. Argh!

Like the European starling, the English sparrow was escorted to these shores by an optimistic human. The bird was supposed to control a cankerworm, as it does in Europe. But as every city dweller knows, these fearless birds will eat anything from bagel crumbs to baloney. In 150 years they have *cheated* their way across North America. And onward. Like starlings, the sparrows have conquered much of the planet. Rather than protecting farmers from insects, the sparrows stuff themselves with grain.

~

Why do some immigrants hit new territory running? Some successful invaders, such as starlings, who hunt on open ground, already specialized in exploiting a disturbed ecosystem. This served them well when they arrived on a continent being cleared for grazing and farming. Others may blossom in their new home because they managed to give the slip to some of their enemies when they left the Old Country. The black rat, for instance, left most of its parasitic worms at home when it emigrated from Europe. Without all the freeloaders, *Rattus rattus* has more energy to channel into reproduction. This dynamic may prove to be common among invasive species, allowing them to do even better in their new country than they did back home. A survey of twenty-six foreign-invader animals finds that the newcomers are burdened by only half as many different parasites as native mammals are.

The melting pot era isn't bygone, not by a long shot. Just fifty years ago, a southern-African bee was imported to Brazil to breed heat tolerance into European honeybees. Oops! They got away! Now the hybrid bees, heat tolerant and hot-tempered to boot, are making niches for themselves all over the southern United States. In 1988, the zebra mussel invaded the Great Lakes, probably after being dumped with a ship's ballast water. It's now fouling (or biofouling, as biologists like to say) rivers and pipes across much of the eastern United States, from the Great Lakes to New Orleans. And 2002 saw the introduction of the voracious "northern snake-head fish" in a Maryland pond. As long as people have traveled, they've taken invasive animals with them. Now, people are moving more than ever, and we continue to transfer animals from one continent to the next.

Some invaders bring with them a redeeming quality. Starlings may pose a fatal threat to the redheaded woodpecker. But they're also an easy meal for predatory birds like my Sharpie. I had glanced out the dining room window just as the creamy, ginger-speckled bird fanned his striped tail for a landing. His claws clutched a wad of dark feathers. He touched down behind the chokecherry tree, but a few rounds of thrashing brought him into view.

Through the window and a forsythia bush I made out a hooked beak and a round eye. The head dipped briefly, and a small cloud of black feathers blew onto the lawn. He plucked quickly, one foot gripping the prey. He lifted his head after each pluck, looking left and right. After five minutes I saw red flesh at his feet. Now he pecked, raised his head, and gulped as he scanned for danger.

Every few minutes he rested, watchful. He fluffed his feathers. Then resumed. He tugged at a string of intestine. He rotated on the carcass. A chickadee flitted near to rubberneck. Somewhere a jay called, and the hawk listened. After thirty minutes, he was sated. He hopped to a low branch of the oak tree. One foot was blood-stained. He picked a feather from between his toes and spat it into the air. He rubbed his beak clean on an upright branch, then rubbed his cheeks. He shook his feathers. His crop bulged with meat. Then he was gone, fueled up for another day.

Even the English sparrows, loathsome as they are, can reveal marvels of natural history. Thanks to them, I know that crows can sniff out concealed food. One day when the sparrow racket was getting on my nerves, I took what was close to hand—large bean-seeds I'd harvested from a past garden—and threw a fistful into the bushes. The sparrows scattered and peace was restored. A couple of hours later the crows came by for a kibble check. I soon realized that one crow was prospecting deep in the fallen leaves and shrubbery. Only once before had I seen a crow venture under bushes. I perked up. The crow walked deeper into the shadows, poking among leaves with its beak. And then it bounded back into the sunlight with a prize: a bean. He pecked it apart and ate it. Half a dozen times he returned to the underbrush to dig out the seeds.

As for invasive plants, I may already have found occasion to mention my hatred for bamboo and Asiatic bittersweet. But it wasn't until botanist Josh Royte conducted a Tour of Horror in the yard that I realized how bad the situation was. The bittersweet and bamboo are big, obnoxious, and hence obvious. Less obvious are the small intruders who already claim ninety-nine percent of my yard. And that's in a Freedom Lawn, which gives natives a much better chance than a typical grass lawn does. For Josh, a botanist for the Maine Chapter of The Nature Conservancy, the plant invasion is an act of war. And he's responding with a similar spirit.

"I'm going to invent a DNA blaster for bamboo," he said when he entered my yard earlier in the year and discovered the Bamboo Wilderness next door. "Something helicopter-based. You dial in the DNA for bamboo, or bittersweet, and *zap!* It wipes out the DNA." Josh's work for The Nature Conservancy entails thrashing around in various ecosystems to make sure a plot of land is worth conserving. If it's infested with foreigners, it's too late. Until the DNA blaster is available, he suggests I might try chewing the bamboo into submission. The shoots taste like celery, he says. Then he spots a sumac baby cowering in the shade of the bamboo.

"Wow, it's not giving up!" he says, crouching over the sprout. The magenta fuzz on a green stem has caught his eye. "Isn't that amazing?" It is. New leaves are unfolding in two tones: dark green, edged in purple. It's the most exotic-looking thing in the yard. But it's not exotic, Josh says. It's native.

So is the aster, he notes, moving on around the perimeter. As are the grapevines that hang in the oak tree. Viking explorers didn't name this part of the world Vinland for nothing. But then we get into a bad patch. The Morrow's honeysuckle, imported by gardeners, is fiendishly invasive. The nightshade is foreign. So are the yellow hawkweed flowers, and the quack grass. Reaching through the fence to pluck a leaf, Josh squeezes the stem. White sap comes out. "Norway maple," he says. "Another major invasive." There's an exotic hop vine on the fence. Among my raspberries, which are native, he finds foreign vetch and orache. "Ballast species," he calls them, because they probably arrived accidentally with the dirt ballast in ships.

"A native Saint-John's-wort!" he exclaims. It's a stick of a plant, with mouse-ear leaves. It lacks the antidepressant punch of its European cousin, but it brightens Josh's affect anyway. Heading back toward the house, we find white and red clovers, both introduced.

"I'm not sure we've got more natives than fingers yet," Josh says. "But this could be another." He pulls up a spindly thing with prickly seed-heads and stuffs it into his shirt pocket. The roots bob under his chin. Coming to the white pines I planted as a wind-

209

break, Josh fingers the foliage. I chose the pines because they're native to my ecosystem. But I may have been swindled. "I'm suspicious," he says. "The needles are really long. I'm guessing it was grown in North Carolina and shipped to a nursery in Maine. So it's . . . native-with-a-wince." Less troubling are the native evening primroses I've allowed to multiply by the deck. Of course, they're surrounded by tulips (Central Asia), daylilies (Europe), peonies (China), and irises (Russia).

"Isn't there someplace else we could look?" Josh pleads, looking back over the lawn. "Maybe we could find one more native?" Beside the driveway, he crawls under the forsythia hedge and hits the jackpot. He finds a probably native violet and a probably native sedge, a dark green plant I would have called grass. Into his pocket go a few threads of moss that might qualify. Out front, he buries his head in the privet hedge to examine a large leaf. "Woo-hoo!" he hoots from the depths. "A native red maple! And an American yew! Yeah!" His yellow shirt is speckled with bark, spiderwebs, and root dirt, and he's looking happier. There's one stem of native poverty grass in my front walk. The adjoining flower bed features three stalks of milkweed that I leave for the butterflies. And on the north side of the house Josh finds a six-inch blue spruce. It's not native to Maine, but it's North American, at least. And that's all.

"Perhaps forty nonnatives," Josh concludes. "And possibly fifteen natives. All of them at risk of being destroyed by invasive species. That's the case all over the country. Invasives are the number two threat, actually. Number one is development and sprawl, which eliminates habitat altogether."

As we head indoors to look up the mystery plants, Josh does a double take and marches across the street to a neighbor's lawn. "Purple loosestrife," he says, aiming a finger at a tall, spiked flower stalk. "It gets out of people's gardens and it destroys wetlands." He stops short of tearing out the flowers. And sighs. "Knowing what I know takes some of the fun out of just enjoying the landscape," he says.

Many of the showy plants that run roughshod over North

America were brought here intentionally. And the comparison between, say, Asiatic bittersweet and peonies illustrates an important difference between a successful invasive and a docile garden plant. Peonies are fussy. They don't spread. They sometimes sulk for years, refusing even to bloom. But bittersweet couldn't be easier to please. Give it a millimeter and it'll take a mile.

Asiatic bittersweet is one of the plants that need a disturbed landscape. If a bird should evacuate a bittersweet seed under a forest canopy, that seed won't find enough light to grow. End of story. But if the seed falls in a spot with no overhanging trees— a road shoulder, the edge of a parking lot, or most of my neighborhood—it's off and running. It will carve out new territory for its own seeds by twisting around neighboring plants and choking them. As the years go by, the bittersweet dynasty will spread across the land by throttling tree after tree. Birds will aid the vine by distributing seeds.

Another advantage bittersweet has over peonies is that bittersweet left more of its diseases back home in East Asia. So while peonies spend energy battling the same old fungi and viruses, the bittersweet feels as though a weight has been lifted from its shoulders. Up it goes! Research has found that the average invasive plant outran three quarters of its fungal and viral persecutors when it emigrated to North America.

I can't dig up any evidence that bittersweet poisons the soil against other plants, though that is a strategy of some invasive plants. Spotted knapweed, a pretty, purple thing from Europe, wins territory in North America by releasing a toxin into the soil. It has evolved immunity to its own poison, as have its old neighbors back in the Old Country. North American plants, however, have evolved no defense against this chemical warfare, which erodes their roots. They fall back in defeat, opening more space for the knapweed.

With all its other advantages, bittersweet doesn't really need chemical weapons. It kills by twining. It grows at a speed that's nearly visible. It produces bushels of seeds, which birds adore. Suckered by the beauty of the fruit, people, too, help bittersweet

211

spread, by clipping branches for indoor display, then dumping them outdoors. Bittersweet has what it takes. So does the bamboo, and the Norway maple.

Which gets me thinking about what might happen if I stopped beheading and uprooting the invaders. What would happen if everybody stopped controlling these colonizers and allowed the suburbs to go wild? In ecosystem science, *succession* describes the order in which plants colonize ground that's been cleared by fire or farming. In my yard, a simplified version of succession might go like this, if only native plants were allowed to play: Goldenrod, milkweed, and raspberries would rule first. They'd be shaded out by white pines. Gradually, hardwoods would mix with evergreens to form the climax population.

But that couldn't happen now, not without a human patron. If I quit chopping down the invasives, I think the result would be a bloody competition between Asiatic bittersweet, Norway maples, and Japanese bamboo. Each is specially equipped for the fight. It would be fascinating to watch. Both the maples and the bamboo open their leaves early in the spring to shade out the competitors. Bittersweet leafs out later, so it would need to exploit the edges and openings. Bittersweet would also have to compete with the aggressive roots of the Norway maples, which form a dense mat just under the surface. But bittersweet's roots are equally formidable, spreading ten or twenty feet to suck nutrition from the soil. The bamboo, too, would have to avoid the shade of the maples. But it also has roots that bore great distances underground to push up new shoots. So maples and bamboo would each shade out competitors under their foliage. Around the edges, bittersweet would rise up, climbing the bamboo stalks. Unable to reach the maples directly, the vines would stretch from the top of the bamboo to the low branches of a tree. It would ascend, choking branches as it climbed, until it blanketed the south side of the tree, stealing the sunlight. The tree would die and fall. Bittersweet and bamboo would race to colonize the sunny spot. And perhaps a stalemate would ensue, with bittersweet gaining one year, and bamboo surging back the next.

In my kitchen, Josh and I look up the mystery plants in *Newcomb's Wildflower Guide,* identifying one more native and two invasives. As he leaves, a flock of starlings swirls off the telephone wire and heads to the backyard.

"I'm going to invent something for starlings, too," Josh says. "It could also be helicopter-based. You fly around, and when you find one of those big flocks, you fire it off. It's sponsored by a dog-food company that grinds up the starling meat."

"Nice," I say. "Does it come with an English sparrow attachment?"

❧

"*Lawn* is a four-letter word." Maureen Austin's fashionable toes weave down a bark-mulch trail in a friend's backyard. Not a blade of grass mars the landscape. Maureen, a tall California blonde, flicks her fingers hither and yon. "Bottlebrush. That's a great hummingbird plant. And migrating tanagers love it, too. Fennel is the food plant of the swallowtail butterfly. Buddleia is a butterfly shrub, of course."

Once again I've sallied forth from my yard, this time wandering as far as the hills east of San Diego. For thirty years, the National Wildlife Federation has been urging lawn owners to make their yards animal-friendly. Maureen's town, Alpine, California, was the first to accept the challenge as a community. NWF has certified the entire city as a Backyard Wildlife Habitat. Some two hundred individual lawn owners now participate. Maureen, with her gardener's tan and gritty fingernails, led the way. And now she has channeled the momentum into a job doing . . . well, being Maureen, from what I can gather. She's a garden diva who circulates and educates and organizes for Mother Nature.

I must come clean: I have doubts about the "wildscaping" concept. My concerns are twofold. First, as is evident in my yard, it's easy to attract invasive plants and animals. I don't want to make life any easier for English sparrows. Second, I can't find many studies of wildscaping's effectiveness, and some minor ones I do find suggest it's not a panacea. I'm not ready to act on my doubts. At home, I leave my dead wood and brush at the yard edges for

213

the fungi and the beetles, as instructed. I'm nursing a patch of milkweed in hope that a monarch caterpillar will someday suspend himself there, in gold and green chrysalis. In grape season, I cede all fruit rights to the catbirds and cardinals. While I do believe it helps, I'm uneasy with the notion that putting out food and water and growing dense shrubbery will automatically aid Nature. I think it's going to require a heavier hand than that to beat back the aggressive, invasive species. But I've come to Alpine for a firsthand view of wildscaping.

Maybe it was a bad idea. When I step out of the car, I don't recognize the birdsongs. I haven't a clue which grasses are native and which are invasive. Even the crows seem unfamiliar, with faster wing beats and higher voices. On the phone Maureen had suggested I spend the morning investigating downtown. So I poke around a hotel that has a Backyard Habitat, and a parking lot with a habitat island. The hotel is adrift in flowery landscaping. It smells great. But it doesn't look wild to me. I don't see any birds. The parking-lot island in the asphalt features some sage, grasses, and flowers. It even hides a miniature pond, in which a plastic bag and a detergent-bottle cap marinate. English sparrows squawk. But did I really expect to find hummingbirds in a parking lot?

I meet Maureen at her Habitat Hut shop, right on the main drag of town. Behind the Hut is an urban jungle of trees and shrubs that is, indeed, all a-flitter with insects and birds. Maureen bundles me into her truck to visit a few of her favorite yards.

She came to this project through gardening, she tells me while we drive through town. Her organic herb garden became so famous that she found her days consumed with garden tours. "And people were saying, 'Wow, where did you get all those birds? How do you get these butterflies? I don't have bees in my garden.' I said, 'Man, this is sad, that people have never seen a butterfly in their yard.'" Her mission was born. One Backyard Habitater I spoke to said he'd watched in disbelief as Maureen rallied the town. "I thought, no way could she do it here," the gardener said. "Alpine's funky. We don't interact too much with our neighbors." But a hundred people came to the first meeting that Maureen

214

advertised. And in 1998, Alpine became the first town to win the NWF's townwide certificate.

Maureen and I step out of the truck at Marilyn and Ed Wojdak's spread, and a cacophony of birdsong washes over us. I recognize none of them. "We've had blue-headed grosbeaks, lazuli buntings, western tanagers, a diamond dove," says Marilyn, who moves with that mixture of grace and stiffness that marks the mature gardner. "So many birds, I lose count." As we bob and duck around towering specimens of bird-of-paradise, jasmine, and asparagus fern, my head bonks a series of net bags stuffed with finch seed. Birdbaths glimmer in the shadows and even in the vegetable patch. On a back fence grows a passion vine bearing zany, white-and-maroon flowers. Marilyn and Maureen bend their heads in search of a painted-lady caterpillar to take back to the Habitat Hut's butterfly garden. Marilyn produces one of the spiked, orange-and-purple creatures and hands it to Maureen on a leaf. Maureen coos over the larva and names it Gloria. With Gloria running wild in the backseat of the truck, we roll on.

The next garden is less shady. Lizards spray grit as they whip out of our way. Crows bray in the neighboring pasture. Farther off, peacocks squeal. The gardener is not at home, but Maureen leads me through a miniature orchard, a cactus garden, and a wilderness of wildlife-pleasing flowers and shrubs. At a third garden we're met by Elma Terry, an elderly woman in a lilac pantsuit and amethyst jewelry. We admire her new water feature, a man-made stream that splashes into a pool of goldfish. Again Maureen rattles off the dozens of useful plants she sees in the garden.

"Which butterflies go for the butterfly bush?" Elma asks as we wander in the heat.

"All of them!" Maureen smiles. On the way back to town Maureen takes me to the Viejas Mall, an Indian-run outlet center. A false river runs through the entire site, babbling in and out of pools and off false boulders. Tall grasses wave in the wind, and sculptures of native animals decorate the stone and concrete. It's half a million times nicer than any mall I've ever seen.

Yet, I doubt. The people I've met seem to be, first and foremost, gardeners. The animals that visit their gardens seem like an afterthought, a bonus. The NWF's Habitat Planning Guide weighs on my mind, too. Using pesticides isn't forbidden in a certified Backyard Habitat, only discouraged. Cats aren't required to be kept indoors. Wildscaping for insects rates a mild suggestion, yet even the babies of seed-eating birds have to eat insects to grow strong.

I'm in no position to throw stones. Look at my yard! It's a Freedom Lawn, yes, but it's much too much Freedom Lawn. Of the few shrubs that provide precious cover for birds, the majority are infested with a murderous gang of English sparrows. *Hey! Who do I think I am?*

On a hilltop in Alpine, I'm reminded who I am: I'm evolving into a natives freak.

Don Hohimer, who lives on the hilltop, is another gardener. He's an amateur botanist, completely bananas about plants. But only natives are given the chance to enter his five-acre Eden. And his is the most splendid and vibrant garden I've ever seen. Huge bushes of white poppies sway by the driveway. Scrappy evergreens screen out the neighboring house. A long arbor embroidered with morning glories banishes the sight of the neighbor's driveway. And a broad hillside, reeking of natural resins and perfume, tosses in the wind like an arid Monet painting that has blown to life. Dueling hummingbirds hurtle past my ear. Finches perch on dry flower stalks, picking out seeds. Bees whir, ants stir. A bunny hops across the patch of grass at the top of the hill.

Don can't stand still to let me admire it. "'Scuse me while I get a weed," he mutters, and lurches into the shrubbery to uproot something. Then he heads down a skinny path that winds around his hill. He's a handsome, muscular guy, suited up for biking, and preconditioned with a smear of road rash under one knee. I have the impression he can't wait until I leave, so he can go bouncing out through the land-trust chaparral that carpets the surrounding hills. He's president of that land trust, and in addition to keeping his five-acre garden weed-free, he founded a community garden, and native-plant gardens in the school system where he teaches. And he has a young family.

"Watch your feet," Don says. "The harvester ants will bite ya. I subject myself to the occasional bite in exchange for having the San Diego horned lizard. That's all they eat. Ooh! A native grass!" He charges up the flowery slope to admire a volunteer. California grasses are surprisingly beleaguered. When Spanish cattle ranchers came through, they scattered European grass seeds to improve the grazing. The local grasses all but quit the field.

"I never fertilize. Never, ever," Don says. We've reached the back of the house where some nonnative grapes are being trained onto wire. Don has a small wine obsession, too. But his Syrah are organic, and they share their rows with wildflowers. How about pests? I ask.

"There's such a diversity of birds and other predators that nothing happens that doesn't take care of itself in a week or so." He has no use for pesticides. And scant use for water.

"Just about everything is native, so it can pretty much survive on its own. I have to water the turf in front of the house. I have small kids, that's my excuse for now. And I occasionally water for fire prevention. You don't want things to get too crispy. Whoa, a whole bunch of weeds!"

He makes a few other interventions. He thins the native chamise shrub, whose famously flammable wood can slingshot a wildfire through the suburbs. And he kills native gophers. He sets traps in their holes, and the cute little buggers instinctively try to block the light that enters through the trap. That triggers the snap. Whack. Left to their own devices, the gophers would eat the roots off the gorgeous manzanita bushes. One must choose. "When I catch one, I leave it for the coyotes and foxes," he says apologetically. "But the crows always get it. Oops, I see a mustard."

We've circled back to the colossal poppy bushes. I look over the chemical-free, waterless hillside that smells like heaven and is exploding with critters. All it requires is weeding, and a touch of gophercide. It's ingenious.

It ought to be. It took millions of years to perfect. Plant after plant, insect after insect, various species carved out a living on this part of the planet. The air grew hotter, and new plants evolved. The air grew cooler, and those plants were replaced by

217

newer models. Animals that ate the old plants evolved into animals that ate new ones. The soil evolved, changed by the plants and animals that lived upon it, by the gases and minerals that the rain and air carried through it. At this moment, the plants and animals living on Don's hill are perfectly adapted to the soil. They get exactly as much water as they have evolved to need. They get precisely the amount of sunlight they've come to require. The soil pleases their palate perfectly. What a notion.

❦

What's the problem with foreign species? certain friends ask me when I fume about the invaders. Aren't we a foreign species, too?

One problem is money: Invaders cost more than $123 billion a year in the United States, according to the calculations of Cornell ecologist David Pimentel. Some $29 billion is incurred by foreign weeds invading cropland, which results in herbicide treatments and crop losses. Another $6.5 billion comes from species like hawkweed and knapweed barging into grazing lands and gardens, displacing fodder and food. And that's only the plants. Rats, the Norway and the black, do $19 billion in damage each year, mainly by gobbling grains. Imported insects are good for $7 billion in destruction. The zebra mussel is at $3 billion and spreading fast. Bringing the problem right to the doorstep of grass farmers, Pimentel estimates that foreign insects cost U.S. lawn owners, gardeners, and golfers $1.5 billion every year.

Another problem is biodiversity, whose dollar value can't be calculated. Invasive species are the top threat facing forty-two percent of threatened and endangered species in the United States. A recent survey of the U.S. National Wildlife Refuge System found that eight million acres of the system—intended to preserve native animals—are infested. Almost half a million acres (and growing) of federal wetlands have succumbed to the same purple loosestrife that brightens my neighbor's fence.

What's the big deal about biodiversity? my devil's advocate friends ask. One answer is that complexity is healthy. Science is demonstrating that Nature's intricate ecosystems, jammed with different life-forms, tend to be resistant to disease and catastrophe. It's also showing that we're unable to restore those natural systems once they're out of whack. A man-made wetland may look good, but under the surface there's nothing going on.

Now, a devil might query, what has a diverse prairie or a bubbling wetland done for me lately? Scientists are just turning to the question of how much of our own health and happiness depends on healthy ecosystems. Some impacts are obvious: It's hard to produce food in a world overrun by aggressive plants, and it's hard to hang on to that food if rats and starlings want it, too. Other effects are subtle. For instance, Lyme disease is less likely to hop from an animal to a person in an ecosystem wealthy in small mammals. Another suggestion of diversity's value is that aspirin came from a willow tree, penicillin from a soil mold, and Taxol from the Pacific yew. If we let a few aggressive invaders drive the natives to extinction, we'll never know what curative chemicals they take with them.

The biggest problem with invasives, though, is the hardest to argue. It has to do with how much a person values the way Nature or God or fate set things up. Before we began shifting thousands of plants and animals around the planet, each little ecosystem evolved slowly. The ebb and flow of glaciers, along with the ticking of genetic mutations, the bobbing of driftwood bearing lizards or beetles, and the crash of an occasional macrometeor, these accidents dictated the pace of change. The resulting ecosystems were fantastically complex, with each plant, animal, and bacterium holding an important position.

I, personally, place a sky-high value on the way Nature built the world. I think the homely Saint-John's-wort is equal in value to the peregrine falcon; a leaf mold is as essential as a brook trout. It's too late to protect North America from the global species swap. Some fifty thousand foreign species are already rearranging our ecosystems to suit themselves. Nonetheless, I think we should make every reasonable effort to protect what pockets of Americana remain. But that's just me.

❧

And speaking of me . . . am I an invasive, exotic species myself?

Author David Quammen has called *Homo sapiens* a "weedy species." But we're much more potent than a weed. Weeds have limits. Norway maples can't cut it in the jungle. Starlings can't make it on the tundra. We can. Furthermore, most weeds, be they furred, feathered, or foliated, need a ride, whether on a boat or driftwood or a hurricane wind, to their new territory. *Homo sapiens* just packs a dog and goes. We're a special case. We're the Superinvasive Species.

However, it seems to me that *Homo sapiens* belongs in the native category, based on the fact that people have been part of North American ecosystems for between thirteen thousand and thirty thousand years. I, personally, whose ancestors arrived only 350 years ago, am a funny question. "Native-with-a-wince," Josh Royte might say: Although I can cross-pollinate with the earlier installments of native *Homo sapiens,* my foliage betrays recent importation. But generally, people have been part of the North American plan for as long as woolly mammoths and saber-toothed tigers have been erased from it. That's native enough for me.

Ditto for dogs, I suppose. Perhaps not ditto for cats, whom Europeans only recently carried to this continent, along with the invasive grasses, bittersweet, English sparrows, and apple trees.

It's swampy question, who belongs here and who doesn't. When I wrangle with friends, I end up arguing for what is practical, as opposed to what is ideal. Practically speaking, *Homo sapiens* has so thoroughly stirred the global pot of species that there's no avoiding our responsibility for the current arrangement. And that responsibility now entails deciding what stays and what goes, what is treasured and what is too much trouble to save. In the Democratic State of Ecology, the entire populace decides how to arrange Nature. It's one person, one vote. And bluebirds are not welcome at the polls.

12

STRANGE FAMILY

STANDING IN SNOW at the back of the yard, I regard the only part of my yard I have yet to examine for wild birds, beasts, and plants: my house. It's a 1917 bungalow, built in an era when people were flowing out of the cities and spreading themselves more thinly across the landscape. Each family commandeered a greater patch of the earth's surface. In an apartment building in Portland, a 1917 family might have had one thousand square feet of living space, sharing one tenth of an acre of land on which their building stood. The family in my house expanded into thirteen hundred square feet. They alone had the use of two tenths of an acre, ten times more yard per family than they'd had in the city.

From the back of my yard, the house looks like an island in Nature, a vinyl-wrapped box with lines too straight to be Nature-made. But no house, least of all one under my housekeeping regimen, is an island. Since mosquitoes aren't a problem in the daytime, I habitually leave my back door open on summer days, which is how Cheeky the Chipmunk, multiple bumblebees, and even bagworms arrived in the kitchen. Even without the convenience of an open door, Nature comes in. Too much of it, in the case of spiders and mice, and a truly biblical quantity of it in the case of sow bugs. That's in addition to the microscopic residents of my house dust, who eat, breed, kill, and are killed without ever setting foot outside. Exploiting the outer skin of my house are more critters, including the house finches who nested in a hanging plant on the porch this spring. That ended badly. So did the efforts of the wasps, although they put up a good fight. The house's impact on Nature reaches even beyond its own walls,

221

throwing rings of influence into the local habitat, and out into the global environment. I slog forward through the snow, into the house's shadow, then into its protection. I'll start a census at the core and work my way outward.

At least one more mammal is in my house than it was meant to accommodate. Since November I've heard this animal at night, skittering through the living room ceiling, scuffling in the bedroom wall. In the basement I've found caves excavated in the fiberglass insulation that I stuffed into the bigger holes. I'm grateful that the invader hasn't raided the kitchen. Still, whoever it is has got to go. I have my suspicions as to the identity. It sounds too small for a gray squirrel, and I've never seen a red squirrel in the yard. Besides, neither is nocturnal. Chipmunks rarely move into houses, and they, too, sleep at night. Mice don't.

At the hardware store I find mouse-size live traps. They look plausible, if barely. I buy two. Per directions, I flick a lump of peanut butter into the back, cock the trigger, and set each trap gently along the wall in the basement. (Mice, like many wild animals, travel along edges, where they can move easily but have quick access to cover.) As for what I'll do with my captives, I remember how Chuck Lubelczyk handled the mice he caught in the backyard last summer. He dumped the trap into a plastic bag, then restrained the mouse through the bag until he could grasp its tail. I can do that, then try to determine which species I'm dealing with. Well, I could do that if I had his traps, which, unlike mine, work. When I check my traps in the morning, it looks as though they've been kicked, punched, and bitten. One was mouse-handled until the peanut butter fell out. The other triggered prematurely, and the disappointed customer tried to gnaw his way in. I junk them and revert to the reliable snap-trap. Later, I'll read that when you live-trap mice, you have to transport them to a different solar system if you don't want them to hop home again.

Down here in the cellar, I've grown accustomed to the crackle of sow bugs underfoot. The first year I spent in this house I was distressed by the army of isopods that emerged from the stone foundation in the spring and marched over the floor. I dread squishing bugs. And here were platoons of them, scudding across the cement like quarter-inch tanks. There was no discipline to the army's maneuvers. Some slid east to west, others slid north to south, and many found their way into stored teacups and baskets, saucepans and toolboxes. The whisper of their corpses has become the sound track for retrieving anything from the basement. The first year I tried shop-vaccing them so I could walk to the washing machine without mashing a dozen. But as fast as I collected them, more appeared. Sow bugs can't survive freezing temperatures, so they have to spend the winter in a protected environment, such as the crevices of my foundation. Given the number of isopods on the floor, this knowledge gave me concern about the permeability of my foundation. It also made me wonder why the isopods don't just turn around in the spring and go out the way they came. Instead they surge forward, where I suspect that those who aren't stepped on suffocate. Isopods are fundamentally sea creatures, with gills and a powerful need for moisture. The best theory I can invent to explain the death march is that they're programmed to move toward warmth in the spring, and the inside of the basement is warmer than the outside. I've surrendered to the army now, and I don't try to avoid them. They die and dry out quickly anyway, becoming more crisp than squishy.

Also down here in the basement are a handful of plants who crowd the windows in summer like flowers in a hothouse. This gives me additional concern about the integrity of my foundation. One of the plants is a variegated border plant who lives in a flower garden on the other side of the window. So its roots must weave through the stone and mortar to establish new shoots inside the house. Another is a wild rose, and I have no idea where that came from. On the north side, where only weeds and moss grow outside the windows, it's weeds and moss who find their way in. Green velvet creeps over the masonry windowsill. It's beautiful, and probably happy to lead such a sheltered life. A bittersweet vine, bane of

223

my existence, has breached the window by the stairs. I expect its botanical mandate is to ascend and strangle me in my bed.

Among the other cellar citizens, the fungi are most prominent. This time of year, they're dormant, resting in the spore phase. But when the earth's moisture suffuses the basement in a few months, a fungal frenzy will ensue. Under every cardboard box, colonies will thrive. Thready hyphae will probe through the cardboard and up into the stored books, feeling their way between pages, devouring pulp and print. Fruiting bodies will release spores into the cellar air, and fresh empires will bloom on the backpack, the wet suit, the old rocking chair. The odor of fungal exhalations and decomposing fungi will rise and form a cloud that waits behind the door to the kitchen.

It's no fun conducting a census of the basement. It's dim, obviously it's about as energy efficient as a tent, and I'm reminded for the zillionth time that there's no room to move my books upstairs. Bungalows, built in a simpler time, are a storage nightmare in the Material Age. And with regard to the leakiness, it occurs to me that mice are getting in through holes substantially larger than those required by sow bugs. True, mice can squirt themselves through a gap about the width of a ballpoint pen. But cold air can squeeze through even easier. Wishing my new mousetraps luck, I flee.

Upstairs, the population of houseflies, spiders, and assorted insects has diminished along with the temperature. The flies wedged into the cracks around the windows, hoping like the sow bugs to benefit from my reliance on fossil fuels. The other day I found a ladybug snoring in my sock drawer. Even the fruit flies are gone from the airspace over the compost bucket on the counter. Their larvae are spending the winter outside, under a few inches of soil. Come spring, they won't need an open door to resume their vigil. The scent of alcohol released by decomposing fruit will lead them right through the window screens.

The pantry moths, too, are blessedly absent. I think I destroyed their habitat, but I understand they sometimes lie low for the winter, even indoors. They could flutter back into my life after a few months. I've never dealt with them before this autumn, and I hope never to encounter them again. Their eggs or larvae came home in an infested grocery item. Once hatched, they must have battered around the cabinet at night, laying as many as four hundred minuscule eggs apiece. It wasn't until adult moths constituted an indoor snow flurry that I realized something had gone awry with the provisions. By then, the larvae had penetrated everything. A factory-sealed bag of pecans, when I opened it, was a shivering, webby mass of nuts, worms, and their chaff. A twist-tied bag of couscous featured some couscouses that squirmed. A screw-top jar of cornmeal, when shaken, revealed clumps of meal bound with white silk. Gak. Everything went outdoors to the compost bin. I scrubbed the cabinets and grabbed the clumsy moths out of the air. And as I replaced flours and grains, the food went in the fridge. No moths have bobbled across my horizon for a few months now, but like flea larvae, these insects can stretch their life cycle in lean times. They could be in there, just waiting for a bag of rice.

The greatest diversity of my housemates probably inhabits the house-dust ecosystem. Having researched the world of microscopic mites and microbes and pseudoscorpions for a previous book, I know their natural history well. Even so, I find them hard to believe in, due to their smallness. Returning my disregard, the plants and animals exploit strips of habitat between the floorboards, expanses of shelter in the rugs and couch cushions, and the cavey innards of the mattresses upstairs. Bacteria and fungi grow like grass on a thin soil composed of my copious skin-sheddings, newspaper fibers, and towel lint. Grazing on the "grass" are dust mites and silverfish. The most famous dust mite also devours my skin, raking spit-softened tissue into a body so simplified it lacks even a head. And preying on the grazers are ruthless carnivores. Predatory mites lurk between rug fibers, waiting for an eyeless, headless mite to trundle past. They stab their quarry, then drink them dry. Between books, or under the table leg, sit crablike pseudoscorpions with mite-crushing claws. It's impossible to construe my house as an island in Nature when there's a Serengeti under my sofa, and a river of mite blood flowing between the banks of the maple boards.

Among all the other oddities that come and go through my house, I'm unable to forget the pearlescent egg I carried indoors in the summer. I had found it attached to a pine needle. It was the size of a sesame seed, and beneath the surface I could see two black eyespots. I hoped to see more under the microscope. But I set it in a jar lid on my office windowsill and got distracted. Next thing I knew, I had propped my feet on the windowsill and flipped the lid. I searched high and low and couldn't find the pearl. Perhaps it hatched and departed through the screen, to pursue its destiny. Perhaps I ran over it with my chair. Maybe it's charting new territory for its species, traveling the innards of my computer. Those black eyes haunt me.

On my first night of serious mousetrapping, I score. I shake the mice off the traps and onto a piece of aluminum foil. Crushed skulls aside, they're winsome. Their fur is gray-brown, with a browner shadow running the length of the spine. Their bellies are snowy. One looks young, its fur silken and flawless. The tiny feet shimmer with fine, white fur. The whiskers are longer than the entire head, black near the nose then bleaching to white. The tail is as long as the mouse, gray-brown on top, white on bottom. The black eyes are large, the gray-pink ears as thin as the petal of a violet. Each animal is two inches long, plus tail.

I try to identify their species. I can rule out *Mus musculus,* the house mouse who has followed *Homo sapiens* around the globe. He's all brown, top and bottom, and has a scaly tail. The other two species of house-loving mice look alike. The deer mouse and the white-footed mouse can be sorted only by a few details, and even those are unreliable, I read in an animal-diversity database. I run through the details.

The tail of the deer mouse is supposed to be cleanly divided between the gray top and the white underside; the white-footed's tail is more muddled. My mice have ruler-straight tail lines. One point for deer mouse.

The deer mouse may have hind feet shorter than twenty-two millimeters; the white-footed's foot may be longer than twenty-two millimeters. I take out my ruler. These feet measure twenty millimeters. Another point for deer mouse.

Finally, the deer mouse is more richly colored brown, while the white-footed is more pinky-buff. The color of my mice is difficult to describe. Milk-chocolate truffle? I would say they're rich. These mice are as gorgeously pelted as any lion. A third point for deer mouse.

One of the most amusing facts I stumble across in my mouse-sleuthing concerns the deer mouse's ability to find his way home when dumped out of a live trap. Scientists have trapped and tagged mice and dropped them off far from home. They report that a mile is not far enough. A river is an insufficient barrier. A paradisal new habitat is not adequately attractive. Even relocating the same mouse twice, or thrice, won't discourage it. Scientists found that they could trap a mouse and dump him a half mile away for three days in a row, with the critter returning each following day. Equally curious is that an entire family of these animals will walk into mousetraps, night after night. I placed two traps a foot apart along the basement wall, yet the vision of a dead relative in one trap didn't scare off the second mouse.

Not so amusing is the rodents' role in disease transmission. Both deer mice and white-footeds can carry hantavirus, which causes a dire respiratory syndrome. Their feces and pee can transmit the virus to me. But that's more a concern with the dust of dried waste, which can be inhaled. I'm not much worried. These mice on my desk are fresh and moist, and I'll wash well when I'm done with them. If my diagnosis is wrong and these are white-footed mice, then they're an important host for the deer tick, who transmits Lyme disease. But I bear the local mice no ill will on those grounds. We all carry plagues.

When I'm finished with the mice, it's dark, and the crows have already flown across the harbor to their winter rookery. I fold the mice in the foil, wrap them in a plastic bag, and pop them in the fridge. I spread more peanut butter on the traps and set them back in the cellar.

In the morning I toss the refrigerated carcasses out onto the snow for the crows. Their response is uncharacteristic. Normally, the crows approach my offerings on foot, with caution. Not this time. Two birds dive from the oak tree and barely touch down as they snap up the mice. They beeline to an eating tree. That night, I catch another mouse. And that is the end. This deep into the winter, mice have chosen their nests and arranged their food stores. My walls and ceilings should be quiet until next fall.

The second ring of wildlife habitat my house offers is its exterior skin. I'm fortunate that it is seldom targeted for serious exploitation. Termites, for instance, find these latitudes too cold. Woodpeckers, who have taken to battering holes in my mother's wooden house, have limited their attentions here to the trees and, in winter, the suet bag. I do maintain constant vigilance against my squirrel friends. And the wasps, whose paper cones I initially welcomed on the chance that they were important pollinators, eventually come to represent an energy-efficiency obstacle. In the fall, when a contractor came to discuss blowing insulation into my windy walls, he promised I'd get none of his cubed fiberglass until the wasps were gone.

They put me through the ringer, those wasps. Determined to avoid blasting toxic chemicals around my home, I nearly burned it down instead. The first colony I confronted was a baseball-size nest on the porch ceiling. Recalling that beekeepers use smoke to pacify honeybees, I rolled up a newspaper and found a book of matches. Then, in case a few wasps came home after I applied the sedation, I put on gloves and a Gore-Tex jacket with the hood cinched around my face. I got a bread bag that I would slip over the slumbering village to pull it down.

228

Out on the porch, I lit the paper and climbed onto a chair. The burning newspaper did release smoke up toward the hive. Yes indeedy, smoke wafted right up into the little hole where the yellow-and-black heads were looking at me. But pieces of the burning newspaper also came off my torch. They drifted around the porch. I didn't pay strict attention to where they settled because the smoke was starting to have an effect on the wasps. If I had to describe the change in their attitude, I wouldn't say "sedated" so much as "howlin' mad." When I jumped off my chair and ran for the door, I was surprised to see little fires all over the porch. I stomped out the flames, then banged inside.

When I poked my head out a few minutes later, the hive looked quiet. I was done messing around. I opened my bread bag, hopped onto the chair, sleeved the bag around the hive, and yanked. It was too easy. A series of gray rings, like ripples in a pond, remained on the wooden ceiling. In my hand the nest vibrated. This vibration aroused mixed feelings. Yes, I'd triumphed over my adversary. But these animals, who meant only to feed their baby sisters and produce a new queen, were going to die in a plastic bag in my plastic garbage can. It was undignified. It was not a worthy end for these Amazon warriors. But I wasn't letting them out. I knotted the plastic. As I stood there feeling bad, a few wasps came home to the rings on the ceiling. By nightfall a dozen had gathered, and they packed together for warmth. Sigh.

That was the easy nest. Out back, another clan had found a dozen gaps in the caulking of the eaves, and these they used as tunnels to an interior nest. I climbed a ladder for a better look. The traffic didn't seem heavy, but when I pulled off the aluminum vent and peeked into the eave, I discovered a nest the size of an Anasazi pueblo in the gloom. That was a lot of cellulose. Maybe, all nests considered, I had more insulation than I thought. Nonetheless, I crammed bits of aluminum foil into the gaps in the caulking outside. I did yoga breathing and thought pure thoughts when some returning wasps hummed in for a landing on my fingers. My theory was that they wouldn't be able to chew through the foil, and the colony inside would starve. The guilt caused my pulse to increase, but I concentrated on my breathing. I needed that cubed fiberglass. The overheating planet needed that fiberglass. It would be for the good of all wasps. My own wasps tapped and shoved at the foil. A dozen milled around. Then two dozen. At three dozen, I backed down. Okay, that wasn't so bad.

One to go. This tough nut was swelling to football size at the peak of the roof. My first plan was to fully extend my ladder some dark night when the wasps were dreaming, then climb twenty-five feet to execute a plastic-bag snatch. But my friends vetoed this in favor of my not breaking my neck in the night. I could think of no way around the spray can of poison. The can, of course, warned against inhaling within fifty miles of its contents, or allowing the stuff to touch skin, or letting pets get near it, or allowing butterflies, bumblebees, little Greenies, rove beetles, pearly pine bugs, ladybugs, ants, or Babbette the spider to smell even the faintest concentrations, on penalty of a writhing and painful death. It was apparently extra bad for fish, many of whom live in the ocean a few hundred yards downhill. I meditated to focus my mind: "Mmmmm, fiberglasssssssss."

Chemical salvation was a disappointment. For starters, the wide bungalow eaves kept the ladder at such a flat angle that midway up it became too bouncy to continue. I was fixing to poison the world *and* break my neck. I aimed carefully as the ladder shimmied. A foamy jet spattered on the paper hive. Some of it penetrated the nickel-size hole. Much more bounced and showered down on my upturned face. My hands dampened in the rain. Droplets moistened the lawn and life below. A few wasp corpses, also wet, tumbled past me.

But other wasps seemed immune. The next day they were busily tending the nest. I blasted them again, recoating myself. The label instructed me to bathe immediately. The wasps lived on. The third day I dosed myself a third time—and gave up. I got a broomstick and wobbled close enough to swing at the nest. Naturally, it took about fifteen whacks to dislodge it. But only a few warriors followed me when I clanked back down the ladder. On the walkway below, the fallen hive had split open. One larva, a large green gal, was humping away toward the grass. A dozen more larvae poked heads out of the paper honeycomb that was protected inside layers of wrapping. It was an astonishment, this paper city. I put the outer layers in a plastic bag and left it on the porch for a few days to off-gas. And I bathed.

The hive paper was gorgeous stuff. The various plants that the wasps had chewed up had been regurgitated in stripes of silver, white, gray, muted green, and bronze. It was a winter dunescape. It was a watercolor of endless hills and streams on an overcast day. As homes go, it was stunning. I wished they had built it in a tree, but as a habitat choice, my house probably offered superior protection from the elements.

I had the same thought when a pair of house finches set up housekeeping on the porch earlier in the year. Unlike a tree canopy, the porch roof is leak-proof. The first time I startled the female, I thought the hanging pot of fuchsia was an odd place for her to be insect-hunting. After surprising her there a half dozen times, I had to see these irresistible insects for myself. Climbing onto the railing, I found a nest fitted among the fuchsia stems. An egg the color of a milky sky rested in the bottom. Mr. and Mrs. Birdlesman, he wearing a handsome red hood, sat on the phone wire chirping until I left.

I started greeting Mrs. Birdlesman each time I came and went. She began to stay put instead of flying off. When five eggs clustered in the cup, she settled in to incubate. The day I saw her return to the nest with food in her finchy beak, I knew she had succeeded. After she departed, I peeked. Four purple, heaving birdlets with yellow beaks lay in the cup, along with a dud egg that would disappear in a couple of days. The Babies Birdlesman grew. They propped themselves up. They squalled to beat the band. (I recently read that when chicks coordinate their yelling, their parents work harder.) The Birdlesmans made more and more trips with food. As the kids swelled and fledged with rough feathers, their feeding became constant . . . and someone noticed. Or maybe someone noticed me, when I hopped up on the railing to see them every day. Either way, the location of the nest was discovered. One afternoon the parents sat chirping together on the phone wire. They hadn't had time for that in weeks.

I climbed on the railing. One kid remained, dead. I blame a blue jay, who could have swallowed two whole and carried a third away. A crow would have battered leaves off the fuchsia, I think. I doubt a squirrel could carry three fledglings or chew them up at the scene without leaving blood. A cat might have jumped to the planter, but like the crow would have done obvious damage.

The Birdlesmans didn't make the same mistake twice. They renested somewhere in the backyard, continuing to visit the thistle-seed feeder. And by late summer they were escorting a handsome brood.

Babbette the spider and her front-porch sister both had better luck with the human habitat. Each enjoyed waterproof shelter when foul weather struck, whereas a third sister inhabiting the flower garden spent rainy days under a leaf. This back-to-the-land sister blended well with her environment, but was nonetheless more vulnerable to birds than were Babbette, who could retreat into the eave vent, and the porch sister, who often waited for her meals under the porch ceiling. When Babbette first set up shop outside my window, I worried that wind eddies around the house might deprive her of insects, but she became the fattest of the three.

The other folks who benefit from the outer skin of my house are usually short-term visitors. The seagulls normally patronize the tall chimneys on a house a block away. They changed their routine when I was working out the kinks of crow-feeding. It took me a while to learn that any morsel larger than a nickel would attract the seagulls. So for a while, the gulls moved to my roof, waiting silently for me to toss food into the yard. Now that I've settled on small dog kibbles for my crows, the seagulls have returned to their chimneys, which offer a better view of the neighborhood.

The crows sometimes use my house for a perch, too. But it's rare. There was one day when all six picked pears from Neighbor Hugh's tree and carried them to his roof to eat them. But I think that was a special occasion. I never saw them do it again. And although a crow once landed on the gable peak outside my office

232

window, I'm not sure he was trying to get my attention. Spoiled crows have been known to tap on windows or fly in front of windshields if their caretakers forget to feed them. My crows have never been that forward. Even on these cold winter mornings, they don't call for food. They just sit on their begging branch and wait for me to notice them.

It's the squirrels I keep the sharpest eye on. Crook and Fluff often travel from the lilac hedge, to the deck roof, to the house roof. It makes me nervous. Perhaps they just see my roof as another element in their elevated trail system. But old sheet-steel patches on my eaves testify to squirrels' eagerness to domesticate in attics. Who can blame them? I'm sure the leaf balls they construct in trees are marvels of engineering, but I wouldn't trade beds with a squirrel. So I don't turn a blind eye to their expeditions onto my roof. I make scary noises. I throw things. I try to act like a cranky neighbor.

I get to wondering how far my house's sphere of influence extends. I know it extends farther than it ought to. The cap is insulated, but the walls are not. That I confirmed when I rewired the place and found the walls empty. Vacant. Airy. Given the age of the house, it isn't surprising. Just horrifying. I was an impressionable teenager during the energy crisis of the 1970s. My parents insulated the attic of our ancient farmhouse, bought woodstoves, and yelped in pain whenever they found a lightbulb blazing unwitnessed. Later on, the realization that energy produces pollution and changes the climate hit me hard. I developed my own allergic reaction to wasted watts. I became a bloodhound who bayed at the scent of cold air entering around windows. When I bought this old bungalow, I knew I was in for a weatherizing marathon. But I wanted to do it in a rational way, plugging the biggest holes first. Well, no. First I wanted to update the wiring and put in a kitchen. Then, once winter came, I called Wes Riley, energy auditor. It was horrible.

Wes himself is not the least bit horrible. He has a little pony-tail to compensate for some thinning elsewhere, and he makes jokes about his prior drug use necessitating a reliance on a camera rather than memory. It's what he shows me about my house that's horrible. Well, some of it is horrible. Other things he shows me save me thousands of dollars. But it will be a while before I forget the stench of the crawl-space dust that whistled through the living room when he fired up his portable wind.

"We'll start at the top and work our way down," Wes says, scooting through the crawl-space door in a bedroom. The entire upstairs is composed of slanted ceilings that meet knee walls. Behind the knee walls are enormous crawl spaces, vented to let the wind whisk through. "Well, you've got five inches of insulation in the crawl-space floor," he says with some surprise. "But you want to shoot for twelve to sixteen. And you want to eliminate these transitions." He means the slanted ceiling should continue right to the exterior wall of the house, rather than stopping at the knee wall. Heat departs at those transitions, he say. A house wants to mimic the shape of an animal in winter: Roundness minimizes surface area. And it's the surfaces that leak heat.

He emerges and turns his attention to the eighty-year-old windows. I tremble. I know they're dismal. The frames are so rotten I've gobbed them with silicone caulking to keep the glass from falling on the lawn before I can muster the clams to buy replacements. Wes doesn't look nearly as disgusted as I expect. He shakes his head. "Replacement windows are the big lie," he says, rapping on the glass. "What you've got here is R-1. A brand-new, double-hung window would be R-3. That's a very small increase. But here," he says, turning back to the crawl space, "you could go to R-19, and it's a much larger area." If you look at weatherizing in terms of how long it takes for an improvement to pay for itself with saved fuel, he says, replacement windows stink. They take twenty-five to forty years to pay. Window quilts are similarly impractical. I'm stunned. He fingers my curtains—sheers under a heavier drape. "Two layers of window coverings are pretty close to replacement windows," he says. I'm stunned and delirious! I've just saved three thousand dollars on new windows! The glass will fall out someday and I'll have to do something, but for now I'm off the hook.

Wes finds the mold. It's in the cold corners of the bedroom, growing up the walls. "Warm, moist air comes up from downstairs, and it goes toward the windows," he says. "Then it condenses on the cold spots." Then mold grows in the wall dew. Eww. Wes's point is that a house functions like a chimney. Heat rises up through it, seeking a way out. This draws in cold air, through a million little leaks. If you accelerate the rising of heat (by ignoring leaks upstairs, or opening a window), then you speed the sucking in of cold air. On the ceiling he finds just such leaks zigzagging across the plaster. "All those cracks put together could make a significant hole right in the ceiling," he says, tracing them with his pen. "The same with pipes and wire chases that go through your walls. Seal them. Insulate the walls. Close the envelope of the house."

In the cellar Wes saves me another five grand. "I'll be replacing the furnace," I squeak, rushing forward to block his view. It's an oil furnace, converted to burn gas. The gas company confessed it's only fifty-something percent efficient; the rest of my gas is wasted. I haven't even turned it on this winter because the gas-burning Jøtul fireplace upstairs is much more efficient. But again, Wes is not revolted.

"This furnace isn't that old. The conversion burner makes it very inefficient. Put your oil burner back on," he says, toeing the Beckett burner on the floor. "Restore it to the way it was meant to run. It won't be the most efficient furnace, but with your insulated walls, you're going to use it less." I am again gleeful!

My glee is short-lived. It's time to plug in Wes's wind. He replaces my front door with a red nylon panel. At the bottom is large fan. A computer measures how much air the fan is pushing. By blowing air out of the house, Wes will simulate a thirty-five-mile-per-hour gale hitting the outside walls. Then he'll calculate how quickly I've been sending my heat out into Nature. He turns on the fan. It roars like a jet engine. He turns it higher. And higher. He's having trouble depressurizing my house to the standard, because air is leaking in so fast. That's bad news. It gets worse. We walk around now, to find leaks. They find us. A hundred storms howl through my house. It feels as though every window is open.

Upstairs, the crawl-space door I so carefully backed with fiberglass is leaking cleavers of cold air around every edge. Where the radiator pipe comes through the floor, a ring of wind rises, too. The outlets I rewired and carefully tucked under their fiberglass: They're spitting cold knives. The only good news is that my windows, their panes slimy with silicone and their rattling frames tamped with plastic wrap, are tight. The room is starting to smell like the dust of the crawl spaces. The fan roars on. At the top of the stairs a light-switch plate feels like an open freezer door. Downstairs in the dining room a gale whirls from each drawer in the built-in cabinet. The whole house smells like the crawl spaces now. Wes grins and opens the cellar door a crack. My hair blows back.

"You think there might be some infiltration in the basement?" he asks. I'm sweating, though the indoor thermometer is at fifty-eight and falling. This is torture. My house leaks like a sieve. It's a giant, wooden strainer. I'm burning priceless fossil fuels to heat the great outdoors. My sweat goes clammy in the wind. Wes is talking about the efficiency industry, but I can't concentrate. Heat is leaving my house even as he speaks and I cannot bear it. I walk toward the fan so he'll shut the blasted thing off. He reads the fan's computer. My house is undergoing a full air-exchange every hour. The goal is three hours. Horrible. But at least he turns the fan off.

"Do you think the house was tighter when it was new?" I ask. "Before the plaster cracked and the shingles shrank?"

"No, energy was cheaper," Wes says as he packs up his instrument of torture. "Coal was five bucks a ton." And climate change certainly didn't enter into the heating philosophy. When a winter gale struck, people just shoveled more coal into the furnace.

Curious about the efficiency of suburban houses compared to apartment buildings in the city, I scrounge up Department of Energy statistics. I'm surprised. I expected a hive of apartments to benefit from their shared walls and many ceilings. But the benefit is tiny, in terms of energy per square foot of dwelling. Big apartment buildings and freestanding homes like mine (better insulated than mine, I presume) are the two most efficient housing types, as a national average. More energy hungry are attached single-family homes; mobile homes; and apartment buildings with less than five units—the latter are least efficient, by far. The statistics are complicated, however. Single-family homes are larger, so although they're as efficient as apartments in a big building, they still use more than twice as much total energy. We do. I do. No, I probably use much more than twice the energy, because I'm losing all my hot air every hour. Unbearable!

And my loss is someone else's gain. Are the birds outside my house basking in waves of infrared radiation? I know whom I want to ask. I just don't know if he'll answer.

∿

When I tell friends that Amory Lovins is coming to my house, most of them squint and ask, "Who?" I tell everyone anyway. Once in a while, I encounter someone whose eyes fly open. For those of us who cry over spilled therms and wasted watts, Amory Lovins is a guru. The waste that preoccupies me on a household level, he considers on a planetary and visionary scale. When I was a young editor at an environmental magazine in the early 1980s, I often gathered facts from his Rocky Mountain Institute, an energy think-tank in Colorado. A few years later a feature in *Outside* magazine related a conversation between him and his partner and ex-wife, Hunter. Hunter had joined the local fire department and put one of those mini-license-plates on her truck. Amory had calculated how the increased drag was eroding her gas mileage— by, you know, many inches per gallon. Readers were intended to read this and think, "Who could live with this man?" I read it and sighed, "Maybe one day, someone will calculate *my* drag." These days, Amory Lovins is an energy consultant to the nations of the world, including mine. He's a Major Player. And he's coming to my house. I'm giddy. And guilty. Having Amory Lovins come to my house gives me insight into how a naughty Catholic feels on the way to confession. But in truth, I'm more interested in Amory's big picture than his critique of my own house. I'm interested in the cumulative effect of replacing Nature with houses.

Amory did not come in winter. I took him when I could get him, which was on a sweltering afternoon in August. He was in Maine for another appointment, and his scheduler had scheduled him to visit my house in a spare hour.

When he arrives, he seems not to know quite why he's here. He is, after all, a Major Player. I, by contrast, am not. Letting him in, I'm struck by the contrast between his expertise and his personal conduct. It's about a hundred degrees out, and he's wearing long pants and a long-sleeved shirt, both dark blue and heat-absorbing. Perhaps he's crunched the numbers and determined that the evaporative cooling of the extra sweat is more effective than wearing shorts. Or maybe he's read about the bedouin, who dress in black robes—theory has it that black maximizes the chimney effect of air rushing upward under the robes, cooling their skin. But Amory's belt would thwart the chimney effect, and the belt is essential to hold up his battery of electronic gizmos—phone, camera, whatnot. Three pens peek from his shirt pocket.

Amory Lovins, a compact man with a neat mustache, glances at me, then stares at my windows. "You know how to use venetian blinds!" he exclaims. "A lost art!" I beam and blush. I've got them tilted so that direct light is repelled, but you can still see out. Amory peeks through the slats at the yard, then sits down, snaps open his titanium Apple computer, and directs his brown gaze at me through wire-rim glasses. He's ready.

My house began its climate-changing career the day the cellar was dug, Amory says. The innumerable carcasses of carbon-rich fungi, bacteria, and other little organisms were uprooted and exposed to the air. "Some of those rotted microorganisms were turning into . . . um . . . finely divided coal?" he says. "But instead they were oxidized." The exposed carbon was embraced by oxygen and carried off as carbon dioxide, or CO_2, climate-changer par excellence. That was my house's first act of climate-changing. But houses are a mixed bag of carbon costs and benefits. They release some carbon into the air; and they also fix some, taking it out of circulation.

Amory looks around the room. "When the house was built, you fixed carbon in the wood. But you released carbon from curing cement, and from manufacturing glass and metal." He peers at the ceiling. "If those are cellulosic tiles, there's some carbon stored there." He turns to his computer to rattle up a paper for my reference, but when he spins the screen toward me, he's suppressing a tiny grin under his mustache. Neither a graph nor a bibliography confronts me. It appears that this legendary geek has a soft side, a biophilic side, a monkey-loving side. Before me is a photo of an Indonesian woman grooming two small primates while a third works over the woman's hair.

"Just a little social grooming," Amory observes. "You know"— he reaches over and picks an imaginary parasite from my hair— "'Ah! There's a bug!'" Amory Lovins puts my louse in his mouth. (Amory Lovins groomed me!) Then he gets me the reference he was looking for. And he's ready for the next topic.

"So, heat: It goes out." He laughs. "If it didn't, you wouldn't have a fuel bill. A typical house has a square yard of holes in it. But heat also *radiates* from all the material in your house, at different rates. And it heats the trees, and shrubs, and your neighbor's house."

He tells me a little bit about my specific house: The wide bungalow eaves shade the house in summer when the sun's angle is high. In winter, the sun falls low enough to duck under the eaves and beam in the windows. The southwest side sports huge windows. On sunny winter days, these windows gather nearly enough radiation to heat the downstairs. But on cloudy days and at night, an avalanche of cold air slides down them and washes across the floor. With no curtains or shades drawn, they are, as Wes Riley pointed out, just one R-value better than huge holes in the wall. Amory also approves of my front porch (beam-beam, blush-blush), but not just because it cools summer air around my house. He likes it because when I sit on it, I promote a sense of community. (This is true—I do chat with neighbors as they pass.) And Amory approves of my trees.

They're on the southwest side from whence the sun is most insistent in the summer; and they're deciduous, so that when I need those warm rays in winter, I get them. But he likes their vivaciousness, too. His house-and-institute in the Rockies features an atrium with fruiting banana trees—and an iguana.

"It's nice to junglify your house," Amory says. "Why would you want to keep your relatives out?" Well, I can think of a couple reasons, but maybe spiders and deer mice are the drunken uncles in my family.

Amory's expertise has grown beyond individual houses. In his visionary world, all new houses would be built like his, with a mile of insulation, gas-filled windows, southern orientation, and solar panels on the roof. He uses one tenth as much energy as the rest of us. Imagine how little oil we'd import if every house were so efficient . . . Amory has imagined that sort of thing and has also diversified into imagining a superefficient car he calls the Hypercar. He and a pal have coined a term that sums up their approach. *Negawatts* are those watts you save by superinsulating, then buying a smaller furnace or air conditioner; or by replacing all your light-bulbs with compact fluorescents; or by taking the fire-department plate off your truck. It's a cheaper way to produce new energy than building more power plants, Amory says. Every one of us can generate negawatts in our own homes and flood the market with power.

He stands up. "I'm going to tune your blinds, if you don't mind," he announces. The sun has dropped, and hot light is sneaking through the slats. He goes from one to the next, twisting the wands. "There," he says, looking up at my carbon-storing cellulosic tiles. "Now the light goes up onto the ceiling as God intended."

Months after Amory's visit, now that the backyard is bright with snow and a natural gale is thrashing my heat into the neighbor-hood, I hunker down to calculate my own drag, so to speak. I want to see how much energy I consume, and how many polluting gases and dusts flow forth from my house. This takes three full days and generates a stack of paper an inch high. No two federal agencies collect quite the same data, so I spend hours with a calculator, converting apples to oranges. To be as accurate as possible, I need data from my own (cold) part of the country, and pollution formulas tailored to the local mix of electricity generators (coal, nuclear, hydro, oil). So, three days later, I can characterize the past year's damage with a middling degree of certainty:

I burned 65,400 cubic feet of natural gas and used 3,420 kilowatt hours of electricity. I'm thrilled to report that in both cases my consumption (due to obsessive switch-flipping, and heating only part of the house) is one-third less than the New England average. Still, when I calculate the mass of my by-products, the tonnage is gag-worthy. My mess, some of which wafts from my gas Jøtul and cookstove, and some of which pours from the region's power plants, looks like this:

- 10,264 pounds of CO_2. If that seems like a lot, that's because only the C comes from burning fuel. The two heavy Os hook onto the C in the air, nearly quadrupling its weight.
- 103 pounds of nitrogen oxides. These contribute to smog, acid rain, and algae growth in water bodies.
- 39 pounds of SO_2. This is another acid-rain precursor. It also alters the behavior of clouds, indirectly changing the planet's temperature.
- 3 ounces of large particles (one tenth of a hair's width or smaller). These mess with clouds, squelch visibility, and collect in your nostrils.
- 10 ounces of tiny particles (a twenty fifth of a hair's width or smaller). These are worse than big particles, because they fly deep into animal lungs and kill sixty thousand people a year in the United States. Ten ounces of anything this small is a ton.
- 6 pounds of volatile organic compounds. Some are carcinogens, others are smog-makers.
- 193 pounds of methane. Methane is a greenhouse gas that's twenty-one times more potent than CO_2.

241

Gross. That's upward of five tons of pollutants produced in heating and powering my house for one year.

❧

All the houses on my street are throbbing with heat this time of year. All the office buildings downtown are pulsating therms into the world, too. They're heating up the masonry and the streets, the trees and the soil. They're building heat islands. They're changing the planet.

Heat islands are most troublesome in summer. That's when Atlanta gets so hot that its rising air generates Atlanta-brand thunderstorms. Houston's hot bubble gives it the dubious distinction of hosting more lightning strikes than anywhere outside Tampa Bay, Florida. The heat rising over Houston also sucks in moist air off the Gulf, building clouds that dump a forty-four percent rain bonus on the cities downwind. The excess heat in summer speeds the baking of smog, adding insult to injury. And it drives people to crank up the air-conditioning, which burns more fossil fuels, which produces more heat and smog . . .

But heat islands also persist in winter. The heat island of Barrow, Alaska, appears *only* in winter. At that time of year, researchers speculate, it's the leaking heat from furnaces that push the city's temperature eleven degrees higher than the surrounding tundra. I suspect that summer-heated masonry and asphalt do the most urban-heating in southern cities, while in northern cities the winter conflagration of fuels makes the greater contrast. But this is just a guess. Research focuses on the summer problem, which overheats people, fouls the air, and instigates wild weather.

It's not a guess to say that even my plant-rich neighborhood produces a heat bubble of its own. Whereas a major city can blow a ten-degree bubble, the suburbs may only add a couple of degrees around themselves. But it counts. It adds up. It warms the world.

My contribution to warming and dirtying the planet is already fairly depressing, so I may as well tally up my car, too. Fortunately

242

it's a newish, smallish Toyota, and my daily commute is zero miles. Still, my periodic research trips, grocery shopping, and visiting were good for eight thousand miles this year. So add onto my bill:

- 5,000 pounds of CO_2
- 20 pounds of nitrogen oxides
- 11 pounds of volatile chemicals

Horrors! My total is in the eight-ton range now. Good thing I don't drive a Ford Explorer, which would have doubled my auto tab.

During Amory's visit, I asked him if he knew of research equating a gallon of gasoline to, say, pounds of dinosaurs that decomposed to produce that gas. He didn't. Shortly thereafter, though, a young scientist published just such a paper. Ecologist Jeff Dukes was tooling around Salt Lake City in a university-owned Chevy Suburban when he became curious about the tonnage of ancient plants burning inside the engine. The answer: ninety-eight tons of plants per gallon.

I can't envision what ninety-eight tons of plant material might look like. Dukes says it's forty acres of wheat, but I can't picture forty acres, either. I cast about for other things that weigh ninety-eight tons. That's sixteen elephants. Or about one and a half million carrots. Or, in the spirit of the evergreen season, it's sixty-five hundred seven-foot Christmas trees—to produce a single gallon of gas. It's still hard to get my head around. How about this: In my car, I burn the equivalent of two hundred ancient Christmas trees per mile. It takes one tree to back my car out of a parking space. A Chevy Suburban burns two trees to back out of its space. It's mind-bending, the rate at which we consume the bygone plants that became fossil fuels.

More elusive is the other side of the equation: How much carbon do the plants in my backyard mop up and store? A generous estimate for a tree is twenty-five pounds of CO_2 per year. And that applies to young, healthy trees like my pines and the oak, not my decrepit apple trees. So my "forest" of three young trees is mopping up an underwhelming seventy-five pounds of my house and car carbon. That probably doesn't even compensate for my lawn mower, which I've neglected to consider. I've also left out my airplane travel, and my share of the consumer-goods-and-services network that grows, ships, and refrigerates my food; builds, ships, and heats my hardware; and heats my movie theater and my favorite Chinese restaurant. A quick'n'dirty average for a person living in the United States is an annual output of twenty tons of CO_2. So, if I'm going to rely on backyard trees to clean up after me, I need to plant another . . . sixteen hundred trees. And that doesn't include my methane bill, my tiny-particles bill, my volatile-organics bill, my share of the oil spills and nuclear-waste storage, and a host of other energy-related debts. On the bright side, it also excludes that my soil, grass, and shrubbery are active carbon-soakers—when they're not frozen under a foot and a half of snow.

My debt to the atmosphere is representative of the national situation. The United States is estimated to sport enough greenery to soak up between ten and thirty percent of our carbon emissions. This tally of grubbiness is, at least, reinspiring me to jam plastic wrap into every crevice of my bungalow, and to make a new attempt to consolidate my auto-assisted errands into a couple of trips a week. (Don't even get me started on what Amory Lovins has to say about cars—they use the vast majority of their fuel to move themselves, not me.) Why, oh why, did I buy this house and move out of the city, where the best movies and the best grocer were a fifteen-minute walk away?

Well, here I am. And I'm here for the same reasons that millions of others are: I want an independent house. A piece of green earth. A little distance from the neighbors. And nice neighbors, who share my admiration for greenery and calm.

I got so much more. I got more burden than I bargained for. I didn't foresee how seriously I'd take my stewardship of this rectangle on the planet's surface. Knowing that every pass of my lawn-mower blade changes the lives of animals makes me question why I mow at all. Knowing that my forsythia bushes take up space that could support a native shrub, whose blossoms could feed a native insect, which might sustain a struggling songbird, this gives me pause. Where do I fit, on this one-fifth acre? What are my rights and responsibilities? These are ethical puzzles I didn't anticipate.

On the other hand, I got many more nice neighbors than I dreamed I would. Even now, in the dead of winter, they're out there. I'm happy to see my squirrels, fractious tail-snippers though they may be, going about their work in sun or snow. My insect friends are suspended in the ground, hardly even alive, but still capable of bouncing to life in the spring. Deeper in the earth, my woodchuck is curled in a heat-saving sphere. I'll be so happy to see her broad head plowing through my weeds again.

I even got some loud and difficult neighbors I hadn't planned on. My crows rise early, and they like to shout about it. But a day when their black faces don't turn up to peer in my windows leaves an empty feeling.

Neighbors, friends, the division is blurry. The Indians who owned my yard a few hundred years ago had a word that meant "relatives of a strange race": *ntu'tem*. It referred to a myth in which thirsty people ran into flowing water, where they were transformed into the animals of the world. For the Armouchiquois, the creatures that lived around them were family. Perhaps that's what Cheeky the Chipmunk, Babbette the spider, the crows, even the trees and mushrooms are to me: *ntu'tem*. Strange family, indeed. But my family. Mine to cherish, and mine to take care of, to the best of my ability.

REFERENCES

1: A Flood of Feathers

Able, K. P. 1999. *Gatherings of Angels: Migrating birds and their ecology.* Ithaca: Comstock Books.

Bhattacharjee, Y. 2002. "In the animal kingdom, a new look at female beauty." *New York Times*, June 25, F3.

Blair, R. B. 1996. "Land use and avian species diversity along an urban gradient." *Ecological Applications*, 6 (2): 505–19.

Blount, J. D., et al. 2003. "Carotenoid modulation of immune function and sexual attractiveness in zebra finches." *Science* 300: 125–27.

Browne, M. 1993. "Migrating birds set compasses by sunlight and stars." *New York Times*, September 29, C1, C9.

Caffrey, C. 2000. "Correlates of reproductive success in cooperatively breeding western American crows: If helpers help, it's not by much." *The Condor* 102:333–41

Clark, L., et al. 1985. "Use of nest material as insecticidal and antipathogenic agents by the European starling." *Oecologia* 67:169–76.

———. 1988. "Effect of biologically active plants used as nest material and the derived benefit to starling nestlings." *Oecologia* 77:174–80.

Coleman, J. S., et al. 1997. "Cats and wildlife: A conservation dilemma." University of Wisconsin–Extension. http://wildlife.wisc.edu/extension/catfly3.htm.

Ehrlich, P. R., et al. 1988. *The Birder's Handbook: A field guide to the natural history of North American birds—the essential companion to your field guide.* New York: Simon & Schuster.

Faivre, B., et al. 2003. "Immune activation rapidly mirrored in a secondary sexual trait." *Science* 300:103.

Feare, C. 1984. *The Starling.* New York: Oxford University Press.

Ferguson, H. 2000. "Urban birds: A millenium review and future directions." Urban Wildlife Symposium, Nashville, September 12–16.

Kilham, L. 1989. *The American Crow and Common Raven.* College Station: Texas A&M University Press.

Long, K. 1997. *Hummingbirds: A wildlife handbook.* Boulder: Johnson Books.

Madge, S., et al. 1994. *Crows and Jays.* Boston: Houghton Mifflin Company.

McGraw, K. J., et al. 2001. "The influence of carotenoid acquisition and utilization on the maintenance of species-typical plumage pigmentation in male American goldfinches *(Carduelis tristis)* and Northern cardinals *(Cardinalis cardinalis).*" *Physiological and Biochemical Zoology* 74 (6): 843–52.

Mennill, D., et al. 2002. "Female eavesdropping on male song contests in songbirds." *Science* 296:873.

Morris, G. 1990. "Fall's magical mystery tour." *Southern Living,* n.d., 60.

Tangley, L. 1996. "The case of the missing migrants." *Science* 274:1299–1300.

Wiltschko, W., et al. 2002. "Lateralization of magnetic compass orientation in a migratory bird." *Nature* 419:467–70.

Wolfenbarger, L. L. 1999. "Female mate choice in Northern cardinals *(Cardinalis cardinalis):* Is there a preference for redder males?" *Wilson Bulletin* 111 (1): 76–83.

2: Bugs in My Belfry

Bourke, A. F. G., et al. 1995. *Social Evolution in Ants.* Princeton: Princeton University Press.

Buchmann, S. L., et al. 1996. *The Forgotten Pollinators.* Washington, DC: Island Press.

Centers for Disease Control. 2002. "Acute flaccid paralysis

syndrome associated with West Nile virus infection—
Mississippi and Louisiana, July–August 2002." *MMWR* 51
(37): 825–27.

———. 2002. "Intrauterine West Nile virus infection—New York,
2002." *MMWR* 51 (50): 1–8.

Enserink, M. 2002. "West Nile's surprisingly swift continental
sweep." *Science* 297:1988–89.

Evans, H. E. 1968. *Life on a Little-Known Planet*. New York: E. P.
Dutton & Co.

Foelix, R. F. 1996. *Biology of Spiders*. 2nd ed. New York: Oxford
University Press.

Gosline, J. M., et al. 1999. "The mechanical design of spider silks:
From fibroin sequence to mechanical function." *Journal of
Experimental Biology* 202:3295–3303.

Heinrich, B. 1979. *Bumblebee Economics*. Cambridge, MA: Harvard
University Press.

———. 1996. *The Thermal Warriors*. Cambridge, MA: Harvard
University Press.

Kremen, C., et al. 2002. "Crop pollination from native bees at risk
from agricultural intensification." *Proceedings of the National
Academy of Sciences* 99 (26): 16812–16.

Lenahan, M. 1983. "Notes of a mosquito hunter." *Atlantic Monthly*
June, 61–72.

Malakoff, D. 2002. "Bird advocates fear that West Nile virus could
silence the spring." *Science* 297:1989.

McLean, R. G., et al. 2001. "West Nile virus transmision and
ecology in birds." *Annals of the New York Academy of Sciences*
951:54–57.

Minorsky, P. V. 2003. "The hot and the classic." *Plant Physiology*
132:25–26.

Tew, J. 2002. Department of Entomology, Ohio State University,
personal communication.

Turell, M. J., et al. "Vector competence of North American mosquitoes (Diptera: Culicidae) for West Nile virus." *Journal of Medical
Entomology* 38 (2): 130–34.

Wilson, E. O. 1971. *The Insect Societies*. Cambridge, MA: Belknap.

3: Lawn of Many Waters

Alley, W. M., et al. 2002. "Flow and storage in groundwater systems." *Science* 296:1985–90.

Beuchat, C. 1990. "Body size, medullary thickness, and urine concentrating ability in mammals." *American Journal of Physiology* 258:R298–R308.

Botosaneanu, L., ed. 1986. *Stygofauna Mundi: A faunistic, distributional, and ecological synthesis of the world fauna inhabiting subterranean waters (including the marine interstitial).* Leiden: E. J. Brill/Dr. W. Backhuys.

Boyd, R. A., et al. 2002. "Human-health pharmaceutical compounds in Lake Mead, Nevada and Arizona, and Las Vegas Wash, Nevada, October 2000–August 2001." USGS Open-File Report 02–385.

Breuste, J., et al., eds. 1998. *Urban Ecology.* Berlin: Springer-Verlag.

Gilbert, J., et al. eds. 1994. *Groundwater Ecology.* San Diego: Academic Press.

Green, E. 2003. "Regulators to let maker test chemical levels." *Los Angeles Times,* Home Edition, November 1, A19.

Hayes, T. 2002. "Hermaphroditic, demasculinized frogs after exposure to the herbicide atrazine at low ecologically relevant doses." *Proceedings of the National Academy of Sciences* 99 (8): 5476–80.

Otto, B., et al. 2002. "Paving our way to water shortages: How sprawl aggravates drought." http://www.amrivers.org.

Renner, R. 2002. "Conflict brewing over herbicide's link to frog deformities." *Science* 298:938–39.

Schiermeier, Q. 2003. "Studies assess risks of drugs in water cycle." *Nature* 424:5.

Vitousek, P. M., et al. 1997. "Human Alteration of the Global Nitrogen Cycle: Causes and consequences" (pamphlet). Washington, DC: Ecological Society of America.

Wright, J. M., et al. 2003. "Effect of trihalomethane exposure on fetal development." *Occupational and Environmental Medicine* 60:173–80.

4: I Love You, Now Spit Out My Azalea

Harden, B. 2002. "Deer draw cougars ever eastward." *New York Times* November 12, F1, F4.

Keefe, F. F. 1967. *The World of the Opossum*. Philadelphia: J. B. Lippincott Company.

LoGiudice, K., et al. 2003. "The ecology of infectious disease: Effects of host diversity and community composition on Lyme disease risk." *Proceedings of the National Academy of Sciences*, 100 (2): 567–71.

Steele, M. A., et al. 2001. *North American Tree Squirrels*. Washington, DC: Smithsonian Institution Press.

5: The Army of Earth Movers

Evans, H. E. 1968. *Life on a Little-Known Planet*. New York: E. P. Dutton & Co.

Logan, W. B. 1995. *Dirt: The ecstatic skin of the earth*. New York: Riverhead Books.

McDonnell, M. J., et al. 1997. "Ecosystem processes along an urban-to-rural gradient." *Urban Ecosystems* 1:21–36.

Pouyat, R. V., et al. 1995. "Carbon and nitrogen dynamics in oak stands along an urban-rural gradient." In J. M. Kelly et al., eds., *Carbon Forms and Functions in Forest Soils*. Madison, WI: Soil Science Society of America.

————. 1997. "Litter composition and nitrogen mineralization in oak stands along an urban-rural use gradient." *Urban Ecosystems* 1:117–31.

Steinberg, D. A., et al. 1997. "Earthworm abundance and nitrogen mineralization rates along an urban-rural land use gradient." *Soil Biology and Biochemistry* 29 (3/4): 427–30.

6: The Freedom Lawn

Biser, J. A. 1998. "Really Wild Remedies." *ZooGoer* 27 (1). http://nationalzoo.si.edu/Publications/ZooGoer/1998/1/reallywildremedies.cfm.

Bormann, F. H., et al. 2001. *Redesigning the American Lawn: A search for environmental harmony.* New Haven: Yale University Press.

Casper, B. B., et al. 1997. "Plant competition underground." *Annual Review of Ecology and Systematics* 28:545–70.

Eastman, J. 1992. *The Book of Forest and Thicket: Trees, shrubs, and wildflowers of eastern North America.* Harrisburg: Stackpole Books.

Ehrenfeld, J. G. 2001. "Plant-soil interactions." In *Encyclopedia of Biodiversity.* San Diego: Academic Press.

Gibbons, E. 1962. *Stalking the Wild Asparagus.* New York: Van Rees Press.

Harbor, J., et al. 2002. "Using constructed wetlands to reduce nonpoint source pollution in urban areas" (white paper). www.epa.gov/ORD/WebPubs/nctuw/Harbor2.pdf.

Hector, A., et al. 1999. "Plant diversity and productivity experiments in European grasslands." *Science* 286:1113–27.

Herrick, J. W. 1995. *Iroquois Medical Botany.* Syracuse: Syracuse University Press.

Jackson, R. B., et al. 2002. "Ecosystem carbon loss with woody plant invasion of grasslands." *Nature* 418:623–26.

Jenkins, V. S. 1994. *The Lawn: A history of an American obsession.* Washington: Smithsonian Institution Press.

Manski, D. A., et al. 1981. "Activities of gray squirrels and people in a downtown Washington, D.C., park: Management implications." *Transactions of the North American Wildlife and Natural Resources Conference* 46:439–54.

Nyerges, C. 1999. *Guide to Wild Foods and Useful Plants.* Chicago: Chicago Review Press.

Short, J. R., et al. 1986. "Soils of the mall in Washington, DC: I. Statistical summary of properties." *Soil Science Society of America Journal* 50:699–705.

Tilman, D., et al. 2001. "Diversity and productivity in a long-term grassland experiment." *Science* 294:843–45.

7: Before

Baker, E. W., et al., eds. 1995. *American Beginnings: Exploration, culture, and cartography in the land of Norumbega.* Lincoln: University of Nebraska Press.

Borns, H. W., Jr., et al. 1985. *Late Pleistocene History of Northeastern New England and Adjacent Quebec.* Boulder: Geological Society of America.

Bourque, B. J. 2001. *Twelve Thousand Years: American Indians in Maine.* Lincoln: University of Nebraska Press.

Calloway, C. G., ed. 1991. *Dawnland Encounters: Indians and Europeans in northern New England.* Hanover: University Press of New England.

Drake, S. G. 1835. *The Book of the Indians of North America.* 4th ed. Boston: Antiquarian Insitute.

Fowler, C. M. R. 1990. *The Solid Earth.* New York: Cambridge University Press.

Frenzel, B., et al., eds. 1992. *Atlas of Paleoclimates and Paleoenvironments of the Northern Hemisphere.* Stuttgart: Geographical Research Institute, Hungarian Academy of Sciences, Gustav Fischer Verlag.

Hare, B., et al. 2002. "The domestication of social cognition in dogs." *Science* 298:1634–36.

Jordan, W. B., Jr. 1987. *A History of Cape Elizabeth, Maine.* Bowie, MD: Heritage Books, Inc.

Kehoe, A. B. 2002. *America Before the European Invasions.* London: Longman.

Kelly, J. T., et al. Undated. "Maine's history of sea-level changes." Maine Geological Survey. http://www.state.me.us/doc/nrimc/pubedinf/factsht/marine/sealevel.htm.

Leonard, J. A., et al. 2002. "Ancient DNA evidence for Old World origin of New World dogs." *Science* 298:1613–16.

Maine Public Broadcasting. Undated. "Timeline of Native American Culture." http://www.mainepbs.org/hometsom/timelines/natamtimeline.html#beforemap.

Mlot, C. 1997. "Stalking the ancient dog." *Science News Online.*

http://www.sciencenews.org/sn_arc97/6_28_97/bob1.htm.

Pielou, E. C. 1991. *After the Ice Age: The return of life to glaciated North America*. Chicago: University of Chicago Press.

Piperno, D. R., et al. 2003. "Phytolith evidence for early domestication in southwest Ecuador." *Science* 199:1054–57.

Powers, W. K., et al. 1986. "Putting on the dog: For the Oglala Indians, canine stew is a spiritual delicacy." *Natural History* 2:6–16.

Roerden, C. 1965. *Cape Elizabeth, Maine*. Publisher: Town of Cape Elizabeth.

Sanger, D. In press. "An introduction to the Archaic of the Maritime Peninsula: The view from central Maine." In *The Archaic of the Far Northeast*. Orono: University of Maine Press.

Savolainen, P., et al. 2002. "Genetic evidence for an East Asian origin of domestic dogs." *Science* 298:1610–13.

Scott, C. P. 1995. *Images of America: South Portland and Cape Elizabeth*. Dover, NH: Arcadia Publishing.

South Portland History Committee. 1992. *History of South Portland, Maine*. Privately published.

Swanson, M. T. 1999. "Dextral transpression at the Casco Bay restraining bend, Norumbega fault zone, coastal Maine." Geological Society of America, Special Paper 331.

_____. 2003. University of Southern Maine Geosciences, personal communication.

Wayne, R. K. 1993. "Molecular evolution of the dog family." *Trends in Genetics* 9:218–24.

Whitlock, C., et al. 1997. "Vegetation and climate change in northwest America during the past 125 kyr." *Nature*, 388:57–61.

8: The Stately and Scheming Trees

Archetti, M. 1999. "The origin of autumn colours by coevolution." *Journal of Theoretical Biology* 205 (4): 625–30.

ARS/USDA. 1998. "Signal-sending plants identify their attackers." Press release.

Baldwin, I. T., et al. 2002. "Volatile signaling in plant-plant-herbivore interactions: What is real?" *Current Opinion in Plant Biology* 5 (4): 351–54.

Bennick, A. 2002. "Interaction of plant polyphenols with salivary proteins." *Critical Review of Oral Biology and Medicine* 13 (2): 184–96.

Breuste, J., et al., eds. 1998. *Urban Ecology*. Berlin: Springer-Verlag.

Cadenasso, M. L., et al. 2004. "Effect of landscape boundaries on the flux of nutrients, detritus, and organisms." In G. A. Polis et al., eds., *Food Webs at the Landscape Level*. Chicago: University of Chicago Press.

Coley, P. D., et al. 2001. "Defenses, ecology of." In S. A. Levin, ed., *Encyclopedia of Biodiversity*. San Diego: Academic Press.

Connell, J. H., et al. 2000. "Seedling dynamics over thirty-two years in a tropical rain forest tree." *Ecology* 81:568–84.

Constable, C. P. 1999. "A survey of herbivore-inducible defensive proteins and phytochemicals." In A. A. Agrawal, et al., eds. *Inducible Plant Defenses Against Pathogens and Herbivores: Biochemistry, ecology, and agriculture*. St. Paul: American Phytopathological Society.

Curran, L., et al. 1999. "Impact of El Niño and logging on canopy tree recruitment in Borneo." *Science* 286:2184–88.

Darley-Hill, S., et al. 1981. "Acorn dispersal by the blue jay *(Cyanocitta cristata)*." *Oecologia* 50:231–32.

Farmer, E. E. 2001. "Surface-to-air signals." *Nature* 411:854–56.

Gregg, J. W., et al. 2003. "Urbanization effects on tree growth in the vicinity of New York City." *Nature* 424:183–87.

Hagen, S. B., et al. 2003. "Autumn coloration and herbivore resistance in mountain birch *(Betula pubescens)*." *Ecology Letters* 6:807–11.

Hunter, M. D. 2000. "Mixed signals and cross-talk: Interactions between plants, insect herbivores, and plant pathogens." *Agricultural and Forest Entomology* 2:155–60.

Jones, C. G., et al. 1998. "Chain reactions linking acorns to gypsy moth outbreaks and Lyme disease risk." *Science* 279:1023–26.

Kelly, D., et al. 2003. "Mast seeding in perennial plants: Why, how, where?" *Annual Review of Ecology and Systematics* 33:427–47.

Kessler, A., et al. 2001. "Defensive function of herbivore-induced plant volatile emissions in nature." *Science* 291:2141–44.

_____. 2002. "Plant responses to insect herbivory: The emerging molecular analysis." *Annual Review of Plant Biology* 53:299–328.

Kilgannon, C. 2003. "Get that oak an accountant." *New York Times* May 12, B1.

McDonnell, M. J., et al. 1997. "Ecosystem processes along an urban-to-rural gradient." *Urban Ecosystems* 1:21–36.

McPherson, E. G., et al. 1997. "Quantifying urban forest structure, function, and value: The Chicago urban forest climate project." *Urban Ecosystems* 1:49–61.

_____. 1999. "Benefit-cost analysis of Modesto's municipal urban forest." *Journal of Arboriculture* 25 (5): 235–48.

Piovensan, G., et al. 2001. "Masting behavior in beech: Linking reproduction and climatic variation." *Canadian Journal of Botany* 79:1039–47.

Rowntree, R. A. 1986. "Ecology of the urban forest—introduction to Part II." *Urban Ecology* 9:229–43.

Schnurr, J. L., et al. 2002. "Direct and indirect effects of masting on rodent populations and tree seed survival." *Oikos* 96 (3): 402–10.

Schultz, J. 2003. Schultz Chemical Ecology Lab, Penn State, personal communication.

Stabler, L., et al. 2001. "Effects of mycorrhizal associations on urban tree carbon storage potential." Ecological Society of America, Tucson.

Steele, M. A., et al. 2001. *North American Tree Squirrels.* Washington, DC: Smithsonian Institution Press.

_____. 2002. "Acorn dispersal by birds." In W. J. McShae, et al., eds., *Oak Forest Ecosystems.* Baltimore: Johns Hopkins University Press.

Volk, T. J. 2001. "Fungi." In *Encyclopedia of Biodiversity.* San Diego: Academic Press.

Wojtaszek, P. 1997. "Oxidative burst: An early plant response to pathogen infection." *Biochemical Journal* 322:681–92.

9: Sifting Secrets from the Air

Changnon, S. A., et al. 2002. "A record number of heavy rainstorms in Chicago in 2001." *Transactions of the Illinois State Academy of Science* 95 (2): 73–85.

———. 2003. "Urban effects on freezing rain occurrences." *Journal of Applied Meteorology* 42:863–70.

Hüppop, O., et al. 2003. "North Atlantic oscillation and timing of spring migration in birds." *Proceedings of the Royal Society of London B* 270:233–40.

International Scientific Secretariat (ISS), National Research Center for Environment and Health. 2000. "CAPMAN: Coastal air pollution meteorology and air-sea nutrient exchange. Annual report 2000." Munich: ISS.

Kraft, P., et al. 2000. "Odds and trends: Recent developments in the chemistry of odorants." *Angewandte Chemie International Edition* 39:2980–3010.

———. 2003. "Conception, characterization, and correlation of new marine odorants." *European Journal of Organic Chemistry* 3735–43.

Malakoff, D. 2002. "High-flying science seeks to reduce toll at towers." *Science* 298:357.

Müller, D. G., et al. 1971. "Sex attractant in a brown alga: chemical structure." *Science* 171:815–16.

Nosengo, N. 2003. "Fertilized to death." *Nature* 425:884–95.

Orville, R. E., et al. 2000. "Enhancement of cloud-to-ground lightning over Houston, Texas." *Geophysical Research Letters* 28 (13): 2597–2600.

Parmesan, C., et al. 2003. "A globally coherent fingerprint of climate change impacts across natural systems." *Nature* 421:37–42.

Root, T. L., et al. 2003. "Fingerprints of global warming on wild animals and plants." *Nature* 421:57–60.

Slabbekoorn, H., et al. 2003. "Birds sing at a higher pitch in urban noise." *Nature* 424:267.

Towerkill.com. 2003. Statistics and news regarding bird fatalities at lighted towers.

10: The Thirteen Coldest Days of the Year

Brown, K. 2002. "Surviving the long nights: How plants keep their cool." *Science* 297:1267–68.

Daan, S., et al. 1991. "Warming up for sleep? Ground squirrels sleep during arousals from hibernation." *Neuroscience Letters* 128:265–68.

Gunderson, H. 1976. *Mammalogy.* New York: McGraw-Hill.

Hallman, G. H., et al., eds. 1998. *Temperature Sensitivity in Insects and Application in Integrated Pest Management.* Boulder: Westview Press.

Heinrich, B. 2003. *Winter World: The ingenuity of animal survival.* New York: HarperCollins.

Lambers, H., et al. 1998. *Plant Physiological Ecology.* New York: Springer-Verlag.

Irwin, J. T., et al. 2000. "Mild winter temperatures reduce survival and potential fecundity of the goldenrod gallfly, *Eurosta solidaginis* (Diptera: Tephritidae). *Journal of Insect Physiology* 46:655–61.

Marchand, P. 1987. *Life in the Cold: An introduction to winter ecology.* Hanover: University Press of New England.

Prendergast, B. J., et al. 2002. "Periodic arousal from hibernation is necessary for initiation of immune response in ground squirrels." *American Journal of Physiology* 282 (4): R1054–62.

Steele, M., et al. 2001. *North American Tree Squirrels.* Washington, DC: Smithsonian Institution Press.

11: Melting Pot Blues

Bais, H. P., et al. 2003. "Allelopathy and exotic plant invasion: From molecules and genes to species interactions." *Science* 301:1377–80.

Callaway, R. M., et al. 2004. "Soil biota and exotic plant invasion." *Nature* 427:731–33.

DiMauro, D. 2003. "The effect of urbanization on butterfly species diversity." Ph.D. diss., George Mason University.

Kinzig, A. 2002. "City birds prefer rich neighbors." Ecological Society of America, Tucson, August 8.

Mack, R. N., et al. 2000. "Biotic Invasions: Causes, epidemiology, global consequences, and control" (pamphlet). Washington, DC: Ecological Society of America.

Mitchell, C. E., et al. 2003. "Release of invasive plants from fungal and viral pathogens." *Nature* 421:625–27.

Pimental, D., et al. 2002. *Biological Invasions: Environmental and economic costs of alien plant, animal, and microbe species.* Boca Raton: CRC Press.

Torchin, M. E., et al. 2003. "Introduced species and their missing parasites." *Nature* 421:628–30.

12: Strange Family

Alcoforado, M.-J., et al. 2003. "Nocturnal urban heat island in Lisbon (Portugal): Main features and modelling attempts." Fifth International Conference on Urban Climate, Lodz, Poland, Poster Session 2.

Dukes, J. S. 2003. "Burning buried sunshine: Human consumption of ancient solar energy." *Climatic Change* 61:31–44.

Hinkel, K. M., et al. 2003. "The Barrow urban heat island study: Daily air temperature fields, 2001–2002." Association of American Geographers, New Orleans, March 5–8.

Hungate, B. A., et al. 2003. "Nitrogen and climate change." *Science* 302:1512–13.

Wofsy, S. C. 2001. "Where has all the carbon gone?" *Science* 292:2261–63. (See also "Correction," *Science* 294:2480.)

ACKNOWLEDGMENTS

Science writing is a team effort. Writers, while well equipped with enthusiasm and curiosity, are often clumsy with the nuts and bolts of the disciplines that intrigue them. I leaned heavily on real scientists to produce this book. The time they gave is a great gift, both to me, and to every reader who shares my fascination with how our planet works. My gratitude and admiration fly out to them on crows' wings!

Thanks particularly to Chuck Lubelczyk, field biologist for the Maine Medical Center Research Institute, for biological expertise, and also for many fertile suggestions, connections, and observations that helped shape this book. Thank you, Dave Santoro, weatherman nonpareil. I miss the "weather salons" at the kitchen table! Josh Royte, of the Maine Chapter of The Nature Conservancy, thanks for wading through my alien landscape, and for a thoughtful review of the manuscript. Amory Lovins, energy guru to the world, I still can't believe you made time to visit my humble (and oh-so-inefficient) home! Dr. Tony DeNicola of White Buffalo, thank you for engaging the sharp-clawed question of how people might better share the planet with wildlife, and for reviewing a chunk of verbiage. Dr. David Sanger, anthropologist at the University of Maine, thanks for making the long drive to introduce me to the people who lived in my yard long ago; and for eyeballing my version of the events of the past few hundred thousand years. Thanks also to the members of the Maine Entomological Society who spent a brutally hot day introducing me to my six-legged neighbors. Samantha Langley-Turnbaugh, soil researcher at the University of Southern Maine, thanks for visiting and reviewing. Mark Swanson, geologist at the same institution, thanks for your visit and the maps.

For reviewing great hunks of manuscript, I'm grateful to Dr. Lee Slater, hydrologist from Rutgers; Dr. Katalin Szlavecz, Johns Hopkins champion of soil creatures; and Diana Balmori, urban historian and landscape designer.

I also received the warmest welcome from everyone I met as I traveled to exotic lawns across the nation: Katalin Szlavecz; Dr. Jim Sherald of the National Park Service; Dr. Chris Martin of Arizona State University; Brad and Rod Lancaster, the permaculurists of Tucson; Maureen Austin, the wildlife-gardening goddess of Alpine, California; and the staff at the Institute for Ecosystem Studies.

More thank-yous to the hundreds of other researchers who visited my yard, fielded my questions, and responded with humor, research, and referrals. It's such a pleasure and an honor to work with you!

Even with all these eyes upon it, this book still benefited from the clear vision of editors Panio Gianopoulos and Colin Dickerman at Bloomsbury USA. And despite all *that* guidance, I remain profoundly grateful for the deft navigation of agent Michelle Tessler at Tessler Literary Agency.

I have once again accumulated a huge debt to David Vardeman, Interlibrary Maestro, and his colleagues at the University of Southern Maine library.

And finally, I raise a glass to The Chix, who reminded me to Celebrate Everything. And to Claude, who is wise enough to accept fury as a reasonable response to good criticism, thank you! Again!

A NOTE ON THE AUTHOR

Hannah Holmes is the author of *The Secret Life of Dust*. Her science and travel writing has appeared in publications including the *New York Times Magazine*, *Outside*, *Sierra*, and the *Los Angeles Times Magazine*. She lives in South Portland, Maine.

A NOTE ON THE TYPE

The text of this book is set in Bembo, which was first used in 1495 by the Venetian printer Aldus Manutius for Cardinal Bembo's *De Aetna.* The original types were cut for Manutius by Francesco Griffo. Bembo was one of the types used by Claude Garamond (1480–1561) as a model for his Romain de L'Université, and so it was a forerunner of what became the standard European type for the following two centuries. Its modern form follows the original types and was designed for Monotype in 1929.